Supporting Assessment in
Undergraduate Mathematics

This volume is based upon work supported by the National Science Foundation (NSF) under Grant No. DUE-0127694. Any opinions, findings, and conclusions or recommendations expressed are those of the authors and do not necessarily reflect the views of the NSF.

Library of Congress Catalog Card Number 2005936440

ISBN 0-88385-820-7

Printed in the United States of America

Current printing (last digit):
10 9 8 7 6 5 4 3 2 1

Supporting Assessment in Undergraduate Mathematics

Editor

Lynn Arthur Steen

Case Studies Editors

Bonnie Gold
Laurie Hopkins
Dick Jardine
William A. Marion

Bernard L. Madison, *Project Director*
William E. Haver, *Workshops Director*
Peter Ewell, *Project Evaluator*
Thomas Rishel, *Principal Investigator (2001–02)*
Michael Pearson, *Principal Investigator (2002–05)*

Published and Distributed by
The Mathematical Association of America

Contents

Mathematics-Intensive Programs

Mathematics Programs to Prepare Future Teachers

Undergraduate Major in Mathematics

Introduction

Tensions and Tethers: Assessing Learning in Undergraduate Mathematics

Bernard L. Madison
Department of Mathematics
University of Arkansas
Fayetteville, AR
bmadison@uark.edu

In 2001, after a decade of encouraging and supporting comprehensive assessment of learning in undergraduate mathematics, the Mathematical Association of America (MAA) was well positioned to seize an opportunity for funding from the National Science Foundation (NSF) to intensify and extend this support. As a result, NSF awarded MAA a half-million dollars for a three-year project "Supporting Assessment in Undergraduate Mathematics" (SAUM) that provided a much-needed stimulus for assessment at the departmental level. The need for such a program is rooted in the various and often conflicting views of assessment stemming from worry about uses of the results, difficulties and complexities of the work, and possible conflicts with traditional practices. Faculty navigating through these views to develop effective assessment programs encounter numerous tensions between alternative routes and limiting tethers that restrict options. Against this background the MAA launched SAUM in January 2002.

The goal of SAUM was to encourage and support faculty in designing and implementing effective programs of assessment of student learning in some curricular block of undergraduate mathematics. SAUM leaders were reasonably sure that many faculty would welcome help with assessment because many colleges and universities were under mandates to develop and implement programs to assess student learning—mandates originating in most cases from external entities such as regional accrediting bodies. Our expectations were accurate. We found many faculty willing to tackle assessment but unenthusiastic and even skeptical about the work.

During the three years of SAUM we promoted assessment to hundreds of faculty in professional forums and worked directly with 68 teams of one to five faculty from 66 colleges or universities in SAUM workshops. The final SAUM workshop—restricted to assessing learning in the major—will conclude in January 2006. Most of the 68 teams had two or three members, with two usually attending the workshop sessions. As these teams worked at the face-to-face workshop sessions, as they continued their work back home, and as we promoted assessment to the larger audiences in professional forums, skepticism was evident in lack of enthusiasm and inevitably brought forth arguments against assessment as we were advocating it.

The arguments were basically of two types: tensions and tethers. Tensions are forces that mitigate against meaningful and effective assessment, pulling toward easier and less effective models. A common example is the tension between doing assessment that is effective in plumbing the depths of student understanding and doing assessment that is practical and more superficial. Most tethers are ties to past and present practices that are likely to continue and possibly prevent or restrict developing effective assessment. For example, many instructional programs are tied to traditional in-course testing and have no plans to change, placing significant limits on assessment.

Below, I describe some of these assessment tensions and tethers, along with some ways SAUM tried to ease the tensions and untie the tethers. First, however, I will explore SAUM retrospectively and describe how it evolved from a decade of assessment activity by the MAA. The paper concludes with a more detailed description of SAUM.

From Awareness to Ownership

The SAUM proposal to NSF was based on an unarticulated progression of steps necessary to get college and university faculty fully committed to meaningful and effective assessment of student learning. The first step is *awareness*, the second, *acceptance*; next comes *engagement*, and finally *ownership*.

First, we aimed to make faculty *aware* of the nature and value of assessment by stimulating thought and discussion. Second, we encouraged *acceptance* through knowledgeable and respected plenary speakers at workshops, and collegial interaction with others interested in and sometimes experienced in assessment. Examples of the plenary presentations, documented on the SAUM website,[1] are presentations and writings by Lynn Steen (SAUM senior personnel) and Peter Ewell (SAUM evaluator). Their combined overview of how assessment is positioned in the larger arena of Federal, state, and university policies and practices can be surmised from their article *The Four A's: Accountability, Accreditation, Assessment, and Articulation* (Ewell & Steen, 2003). This article is based on a presentation by Peter Ewell at the face-to-face session of Workshops #1 and #2 at Towson University in January 2003.

Peter Ewell was an unexpected and valuable resource at workshops, giving plenary presentations and generously agreeing to consult with individual teams. His broad historical perspective, vast experience in consulting with and advising colleges and universities, and intimate knowledge of policies of accrediting bodies gave teams both encouragement and helpful advice. Further, Peter's view as a non-mathematician was helpful both for his questioning and his knowledge of other disciplines. Peter's expertise was nicely complemented by Lynn Steen's wide experience with mathematics, mathematics education, and mathematics and science policy issues.

_____
[1] www.maa.org/saum

Third, we urged workshop participants to *engage* in designing and implementing an assessment program at their home institutons. Face-to-face workshop sessions required exit tickets that were plans for actions until the next face-to-face session. Teams presented these plans to their workshop colleagues and then reported at the next session on what had been done. As noted by Peter Ewell in his evaluator's report (pp. 19–26), this strategy provided strong incentive for participants to make progress at their own institution so that they would have something to report at the next session of the workshop.

Finally, we promoted *ownership* by requiring that each team write a case study describing its assessment program or present a paper or poster at a professional meeting. As of this writing, the teams have produced 24 case studies, 24 paper presentations, and 18 posters. Paper sessions were sponsored by SAUM at MathFest 2003 in Boulder, Colorado, and at the 2004 Joint Mathematics Meetings in Phoenix, Arizona. SAUM also sponsored a poster session at Phoenix. Finally, an invited paper session is scheduled for the 2006 Joint Mathematics Meetings in San Antonio, Texas.

Background for SAUM

SAUM's background goes back to an MAA long-range planning meeting in the late 1980s. At that meeting I asked what the MAA was going to do regarding the growing movement on assessment that had entered the US higher education scene about a decade earlier. Indicative of the fact that no plans had been made by MAA, I soon found myself as chair of the 12-member Subcommittee on Assessment of MAA's Committee on the Undergraduate Program in Mathematics (CUPM). We were charged with advising MAA members on policies and procedures for assessment of learning in the undergraduate major for the purpose of program improvement. Very few of the subcommittee members had any experience in or knowledge of the kind of assessment we would eventually understand that we needed, and we struggled with the multiple meanings and connotations of the vocabulary surrounding the assessment movement. Nevertheless, we plowed into our work at the summer meeting in Columbus, Ohio, in 1990.

In retrospect, our work developed in three distinct phases: (1) understanding the assessment landscape that included outspoken opposition to assessment; (2) developing guidelines for assessment; and (3) compiling case studies of assessment programs in mathematics departments. A fourth phase, seen in retrospect, was the extensive faculty awareness and professional development made possible by SAUM.

Two vehicles proved very helpful in Phase 1. First, in 1991, I moderated an e-mail discussion on assessment among fourteen academics (twelve mathematicians and two non-mathematicians) that included four members of the Assessment Subcommittee. Some of the discussants were opposed to assessment as it was then evolving; their worries ranged from operational issues like extra work to fundamental issues like academic freedom. E-mail was neither user-friendly nor regularly read in 1991, and managing the information flow and compiling it into a coherent report was quite challenging. Nonetheless, a report was written and published in *Heeding the Call for Change* (Madison, 1992), edited by Lynn Steen, who had been both helpful and encouraging on my involvement with assessment.

Appended to the report of the 1991 e-mail discussion is a reprint of a seminal article by Grant Wiggins consisting of the text of his 1990 keynote address to the assessment conference of the American Association for Higher Education (Wiggins, 1992). This annual conference began in 1985 and over the past two decades has been the premier convening event on assessment in higher education. Between 1990 and 1995, I attended these conferences, learned about assessment outside mathematics, and eventually mastered the language. Plenary speakers such as Wiggins, Patricia Cross, and Peter Ewell were impressive in their articulate command of such a large academic landscape.

Phase 2 of the work of the Assessment Subcommittee consisted of producing a document on assessment that would both encourage assessment and guide department faculties in their efforts to design and implement assessment programs. Grounded largely in the e-mail discussion and a couple of AAHE assessment conferences, I forged a first draft of guidelines that was based on assessment as a cycle that eventually would have five stages before it repeated. By 1993 the Subcommittee had a draft ready to circulate for comment. Aside from being viewed as simplistic by some because of inattention to research on learning, the guidelines were well received and CUPM approved them in January 1995 (CUPM, 1995).

Further plans of the Subcommittee included gathering case studies as examples to guide others in developing assessment programs. The small number of contributions to two contributed paper sessions that the Subcommittee had sponsored did not bode well for collecting case studies, especially on assessment of learning in the major. However, strong interest and enrollment in mini-courses on assessment indicated that case studies might soon be available. One of the Subcommittee members, William Marion, had expressed interest in teaming up with Bonnie Gold and Sandra Keith to gather and edit case studies on more general assessment of learning in undergraduate mathematics. By agreeing to help these three, I saw the work of the Subcommittee as essentially finished and recommended that we be discharged. The Subcommittee was dissolved, and in 1999 Assessment Practices in Undergraduate Mathematics containing seventy-two case studies was published as MAA Notes No. 49, with Gold, Keith, and Marion as editors (Gold, et al., 1999).

Two years later, in 2001, NSF announced the first solicitation of proposals in the new Assessment of Student Achievement program. During two weeks in May 2001 while I was serving as Visiting Mathematician at MAA, with help and encouragement from Thomas Rishel, and with the encouragement and advice from members of CUPM, most notably William Haver, I wrote the proposal for SAUM. I was fortunate to gather together a team for SAUM that included the principals in MAA's decade of work on assessment: Bonnie Gold, Sandra Keith, William Marion, Lynn Steen, and myself. Good fortune continued when William Haver agreed to direct the SAUM workshops and Peter Ewell agreed to serve as SAUM's evaluator.

In August 2001 I learned that the NSF was likely to fund SAUM for the requested period, January 1, 2002, to December 31, 2004, at the requested budget of $500,000, including a sub-award to the University of Arkansas to fund my role as project director. Because we were reasonably sure of an award, we were able to begin work early and in effect extend the period of the project by several months. The award was made official (DUE 0127694) in fall 2001.

The 1995 CUPM Guidelines on Assessment are reprinted as an appendix to *Assessment Practices in Undergraduate Mathematics* (CUPM, 1995) and an account of MAA's work on assessment is in the foreword (Madison, 1999). Another lighter account of my views on encountering and understanding assessment, "Assessment: The Burden of a Name," can be found on the website of Project Kaleidoscope (Madison, 2002).

Tensions and Tethers

As noted above, throughout SAUM and the MAA's assessment work that preceded SAUM, various tensions and tethers slowed progress and prompted long discussions, some of which were helpful. Some faculty teams were able to ease or circumvent the tensions while others still struggle with the opposing forces. Likewise, some were able to free their program of the restraints of certain tethers, while others developed programs within the range allowed.

A major obstacle to negotiating these tensions and tethers is the lack of documented success stories for assessment programs. Very few programs have gone through the assessment cycle multiple times and used the results to make changes that result in increased student learning. This absence of success stories requires that faculty work on assessment be based on either faith or a sense of duty to satisfy a mandate. In theory, the assessment cycle makes sense, but implementation is fraught with possibilities for difficulties and minimal returns. There is, thus, considerable appeal to yield to tensions—to do less work or to work only within the bounds determined by a tether to traditional practice.

Easing Tensions

The most prominent tension in assessment is *between what is practical and what is effective* in judging student performance and understanding. There are several reasons for this, some of which involve other tensions. Multiple-choice, machine-scored tests are practical but not effective in probing the edges and depths of student understanding or for displaying thought processes or misconceptions. Student interviews and open-ended free-response items appear to be more effective in this probing, but are not practical with large numbers of students. We know too little about what is effective and what the practical methods measure, but we believe that getting students to "think aloud" is revealing of how they learn. Unable to see evidence of value in the hard work of effective assessment, we very often rely on the results of practical methods—believing that we are measuring similar or highly correlated constructs.

To ease this tension between the practical and the "impractical," we recommended that faculty start small and grow effective methods slowly. Interviewing a representative sample of students is revealing; comparing the results of these interviews with the results of practical methods can provide valuable information. Knowing how students learn can inform assessment in an essential and powerful way. We know too little about how mathematical concepts are learned, especially in a developmental fashion, and we know too little about how assessment influences instruction. This is both an impediment to doing assessment and a challenging reason for doing so. One can use it as an excuse for waiting until we know more about learning, or one can move ahead guided by experience but alert to evidence of how learning is occurring and how learning and assessment are interconnected.

Many assessment programs are the result of requirements by accrediting agencies or associations. Often these requirements boil down to applying three or four tools to measure student learning outcomes for majors and for general education. For example, the tools for a major could be a capstone course, exit interviews, and an end-of-program comprehensive examination. Consequently, discipline faculties can meet the requirements by doing minimal work—designating a capstone course, interviewing graduating seniors, and selecting an off-the-shelf major field achievement test—and getting minimal benefits. Reflecting on the results of the assessment and considering responses such as program or advising changes requires more work and raises questions about past practices. This tension *between getting by with minimal work for minimal payoff and probing deeply to expose possibly intractable problems* does offer pause to faculty whose time is easily allocated to other valued work.

The tension pulling toward meeting mandated assessment requirements minimally is reinforced by the bad reputation that assessment has among faculties. This reputation derives both from worries about uses of assessment results for accountability decisions and from numerous reports of badly designed and poorly implemented large-scale high stakes assessments. Unfortunately, few people understand the broad assessment landscape well enough to help faculty understand that their assessment work has educational value that is largely independent of the public issues that are often used to discredit assessment. Fortunately, in SAUM we did have people who understood this landscape and could communicate it to mathematics faculty.

Mathematicians are confident of their disciplinary knowledge and generally agree on the validity of research results. However, their research paradigm of reasoning logically from a set of axioms and prior research results is not the empirical methodology of educational practice where assessment resides. This tension *between ways of knowing* in very different disciplines often generates disagreements that prompt further evidence gathering and caution in drawing inferences from assessment evidence. Eventually, though, decisions have to be made without airtight proof.

This tension is amplified by the complexity of the whole assessment landscape. For example, the so-called three pillars of assessment—observation, interpretation, and cognition—encompass whole disciplines such as psychometrics and learning theory (NRC, 2001).

Assessment of learning in a coherent block of courses often provides information that can be used to compare learning in individual courses or in sections of a single course, and hence to judge course and instructor effectiveness. Such comparisons and judgments create tension *between individual faculty member's academic freedom and the larger interest of programs*. Indeed, learning goals for a block of courses do place restrictions on the content of courses within the block.

Mathematics faculty members are accustomed to formulating learning goals in terms of mathematical knowledge rather than in terms of student performance in using mathematics. This creates tension *between testing what students know and testing for what students can do*. Since judging student performance is usually far more complex than testing for specific content knowledge, this tension is closely related to that between practical versus effective tension discussed above.

Partly because of the nature of mathematical knowledge, many instructional programs have not gathered empirical evidence of what affects student learning. Rather, anecdotal information—often based on many years of experience with hundreds of students—holds sway, indicative of the tension *between a culture of evidence and a culture of anecdotal experience*. Since empirical evidence is often inconclusive, intuition and experience will be valuable, even more so when bolstered by evidence.

Untying Tethers

Mathematics programs in colleges and universities are very tradition-bound, and many of these traditions work against effective assessment of student learning. Sometimes, these tethers can be untied or loosened; sometimes they cannot. The tethers we encountered in SAUM include:

- Tethers to traditional *practices in program evaluations*. We are accustomed to evaluating programs by the quantity of resources attracted to the program—inputs—as opposed to quality of learning outcomes. One reason for this traditional practice is the lack of evidence about learning outcomes, or even an articulation of what they are.

- Tethers to traditional *faculty rewards system*. Traditionally, mathematics faculty rewards are based on accomplishments that do not include educational or empirical research results much less amorphous scholarship on assessment. Even if scholarship on assessment is recognized and rewarded, the outlets for such work are very limited. Unlike the situation in mathematics research, standards for judging empirical assessment work are not widely agreed to and, consequently, are inconsistent.

- Tethers to traditional *in-course testing*. This tether was very apparent in the work of SAUM workshop teams. Going beyond assessing learning in a single course to assessing learning in a block of courses was a major step for many faculty teams. This step involved a range of issues from developing learning goals for the block to logistical arrangements of when and where to test. Even when learning goals were agreed to, assessing areas such as general education or quantitative literacy offered special challenges. Recognizing this tether, Grant Wiggins has compared assessment of quantitative literacy to performance of sports. One can practice and even master all the individual skills of basketball, but the assessment of basketball players is based on performances in actual games. Wiggins concludes that assessment for quantitative literacy threatens all mainstream testing and grading in all disciplines, especially mathematics (Wiggins, 2003).

- Tethers to traditional *lecture-style teaching*. Especially with large classes, lecture-style teaching severely limits assessment options, especially for formative assessment. Some electronic feedback systems allow lecturers to receive information quickly about student understanding of concepts, but probing for the edges of understanding or for misconceptions requires some other scheme such as interviewing a sample of the students.

- Tethers to a traditional *curriculum*. The traditional college mathematics curriculum is based largely in content, so assessment of learning (including learning goals) has been couched in terms of this content. Standardized testing has centered on this content. Students and faculty expect assessment items to address knowledge of this content. Consequently, there is resistance to less specific assessment items, for example, open-ended ill-posed questions.

Components of SAUM

SAUM had five components that were aimed at encouraging faculty to design, develop, and implement meaningful assessment programs. The plan, as outlined earlier, was to move faculty in departments of mathematics from awareness of assessment, to acceptance, to engagement, and finally to ownership.

Component 1. The initial component was aimed at stimulating thought and discussion, thereby raising awareness about assessment and why it could be a valuable part of an instructional program. There were three principal vehicles:

- Panels at national and regional professional meetings.

- Ninety-minute forums at meetings of MAA Sections. Forums were held at seventeen of the twenty-nine sections.

- Distributing the 1999 MAA Notes volume, *Assessment Practices in Undergraduate Mathematics* (Gold *et al.* 1999). At the beginning of the SAUM project a copy of this volume (containing seventy-two case studies) was mailed to the chair of each of the 3000 plus departments of mathematics in two-year and four-year colleges or universities in the United States.

Component 2. The second component involved expanding and updating case studies in *Assessment Practices in Undergraduate Mathematics* and gathering new case studies as the main contents of a new volume. For reasons that are unclear, few of the original case studies were updated. The project had more success in gathering new case studies, mainly because the workshops provided natural vehicles for generating them. Those case studies along with supplementary essays and syntheses constitute the contents of this present volume.

SAUM originally planned to support six areas of assessment:

- The major: Courses in the undergraduate mathematics major, including those for prospective secondary school mathematics teachers.

- General education or quantitative literacy: General education courses in mathematics and statistics, including those aimed at achieving quantitative literacy.

- Mathematics for teachers: Blocks of mathematics courses for prospective elementary or middle school teachers.

- Placement programs or developmental mathematics: School mathematics as preparation for college work.

- Reform courses or other innovations.

- Classroom assessment of learning.

As SAUM developed and workshop teams enrolled, this original list of six areas evolved into five: the major, general education, mathematics for teachers, pre-calculus mathematics, and mathematics in mathematics-intensive majors. Well over half of the sixty-eight SAUM teams worked in just one of these areas—assessment of the major.

Component 3. Development and delivery of the four faculty development workshops plus a self-paced online workshop was the central component of SAUM. As noted above, the workshop teams provided almost all the new case studies and provided a critical audience for selecting resources to support assessment. William Haver was the principal organizer and designer of the SAUM workshops. (He also served as a member of his university's team in the first workshop.)

Preliminary evidence indicates that the four workshops were successful in moving the faculty teams to engagement with assessment and many to ownership. We do not have evidence about the effectiveness of the online workshop. Although the suggested readings in the online workshop are selected to move faculty through the awareness, acceptance, engagement, and ownership sequence, face-to-face support and collegial interaction may be an essential ingredient that is missing from the online approach. Workshop participants repeatedly told us that the interaction among teams was important, and we relied heavily on this feature to move participants from acceptance to engagement and ownership. Knowledge and experience of workshop leaders and presenters seemed to work for awareness and acceptance but not much further.

Although not specified as a goal in the original SAUM proposal, one significant accomplishment of SAUM was identifying and developing leadership in assessment of learning in undergraduate mathematics. SAUM began with six leaders, none of whom claimed broad expertise in assessment or in conducting workshops for faculty on assessment. Since each workshop session would require four or more leaders or consultants, recruiting new leaders seemed essential. We were fortunate that in the first and second workshops several leaders emerged. From these leaders we recruited Rick Vaughn (Paradise Valley Community College), William Martin (North Dakota State University), Laurie Hopkins (Columbia College), Kathy Safford-Ramus (St. Peter's College), and Dick Jardine (Keene State College). These new leaders provided experience in assessment at various levels at a variety of institutions and enriched our subsequent workshop sessions by sharing their experiences and consulting with teams on developing assessment programs. Two of the five—Laurie Hopkins and Dick Jardine, both from the second workshop—assisted with editing of the case studies.

Component 4. Construction of the SAUM website began at the outset of the project. The site, a part of MAA Online[2] has several major components that supported SAUM and continue to provide resources for assessment across the US. These components include:

- An annotated bibliography on assessment drawn from multiple sources. Entries are grouped into four areas: (i) Assessment Web Sites; (ii) Policy and Philosophy in Mathematics Assessment; (iii) Case Studies in Mathematics Assessment; and (iv) Policy and Best Practices in Postsecondary Assessment.

- A communication center for SAUM workshops, sessions at national meetings, and section forums.

- Links to seventy-three sites that have information on assessment relevant to the activities of SAUM.

- A frequently asked questions (FAQ) section containing brief answers to 32 common questions about assessment.

- Online copies of case studies and other papers that were published in *Assessment Practices in Undergraduate Mathematics* (Gold, 1999).

---

[2] www.maa.org/saum

- Postings of new case studies including exhibits and supporting documents. Many of these exhibits and documents will not appear in the print, but will reside on the site to be used as supplements to the printed cases.
- The online assessment workshop.
- The contents of this present volume, upon publication.

Component 5. Dissemination of SAUM employs three media: print (publications and mailing), electronic (development and maintenance of a web site), and personal (presentations at national meetings).

- The SAUM report (this volume) will be offered to over 3000 US mathematics departments in two-year and four-year institutions.
- An extensive overview of the SAUM report will appear as a special supplement in FOCUS.
- The SAUM web site will include the contents of this SAUM report, as well as the several items listed above.
- Presentations at national meetings have so far included 24 contributed papers and 18 poster exhibits at MathFest 2003 and the Joint Mathematics Meetings in January 2004.

Beyond SAUM

Through the SAUM workshops, nearly 200 mathematics faculty members participated in the development and implementation of programs of assessment in 66 college and university mathematics departments. In addition, several hundred other faculty became more aware of the challenges and benefits of assessment through other SAUM activities. The SAUM web site and this volume constitute valuable resources for others interested in assessment.

Nonetheless, the accomplishments of SAUM are probably insufficient to provide a critical mass of experience and understanding to cause assessment to become a natural part of instructional programs in all mathematics departments. Because assessment is largely alien to beliefs of many mathematics faculty and to traditions in most mathematics departments, further work by the community will be needed to overcome the tensions and untie the tethers discussed here. Increased calls for accountability for student learning will keep faculty interested but unenthusiastic about assessment. Only success stories that are documented to the satisfaction of skeptical mathematicians will break through the tacit resistance and cause faculty to take ownership of and work diligently on assessment programs. Perhaps some of the SAUM-inspired programs will provide these stories.

If faculty understood the potential benefits of assessment in increased student learning and how to assess for this

learning, projects like SAUM or an MAA Subcommittee on Assessment should not be necessary. But the chance of this happening is slim. Some three decades ago the MAA's Committee on Placement Examinations anticipated eliminating the need for the MAA Placement Testing Program by educating faculty about placement. The Placement Testing Program was finally discontinued about five years ago—not because it was no longer needed but because the difficulties and complexities of such a program were beyond the MAA's scope of operation. After decades of support from a national program, faculty continue to ask for guidance on placement testing, so MAA is now looking for ways to meet this need. Assessment is much broader and less well defined than placement testing, so support from MAA for assessment in undergraduate mathematics will likely be needed for years to come.

References

National Research Council. *Knowing What Students Know*. Committee on the Foundations of Assessment. James W. Pellegrino, Naomi Chudowsky, and Robert Glaser, eds. Washington, DC: National Academies Press, 2001

Committee on the Undergraduate Program in Mathematics. "Assessment of Student Learning for Improving the Undergraduate Major in Mathematics," *Focus: The Newsletter of the Mathematical Association of America*, 15(3) (June, 1995) pp. 24–28. Reprinted in *Assessment Practices in Undergraduate Mathematics*, Bonnie Gold, et al., eds. Washington, DC: Mathematical Association of America, 1999, p. 279–284. www.maa.org/saum/maanotes49/279.html

Ewell, Peter T. and Lynn A. Steen. "The Four A's: Accountability, Accreditation, Assessment, and Articulation." *Focus: The Newsletter of the Mathematical Association of America*, 23:5 (May/June, 2003) pp. 6–8. www.maa.org/features/fouras.html

Gold, Bonnie, Sandra Z. Keith, and William A. Marion. *Assessment Practices in Undergraduate Mathematics*. MAA Notes No. 49. Washington, DC: Mathematical Association of America, 1999. www.maa.org/saum/maanotes49/index.html

Madison, Bernard L. "Assessment of Undergraduate Mathematics." In *Heeding the Call for Change: Suggestions for Curricular Action*, Lynn A. Steen, editor. Washington, DC: Mathematical Association of America, 1992, pp. 137–149. www.maa.org/saum/articles/ assess_undergrad_math.html

Madison, Bernard L. "Assessment and the MAA." Foreword to *Assessment Practices in Undergraduate Mathematics*, Bonnie Gold, *et al.*, eds. Washington, DC: Mathematical Association of America, 1999, p. 7-8. www.maa.org/saum/maanotes49/7.html

Madison, Bernard L. "Assessment: The Burden of a Name." Project Kaleidoscope, 2002. www.pkal.org/template2.cfm?c_id=360. Also at: www.maa.org/saum/articles/assessmenttheburdenofaname.html.

Wiggins, Grant. "'Get Real!': Assessing for Quantitative Literacy." In *Quantitative Literacy: Why Numeracy Matters for Schools and Colleges,* Bernard L. Madison and Lynn Arthur

Steen, editors. Princeton, NJ: National Council on Education and the Disciplines, 2003, pp. 121–143. www.maa.org/saum/articles/wigginsbiotwocol.htm

Wiggins, Grant. "Toward Assessment Worthy of the Liberal Arts." In *Heeding the Call for Change: Suggestions for Curricular Action*, Lynn A. Steen, editor. Washington, DC: Mathematical Association of America, 1992, pp. 150–162. www.maa.org/saum/articles/wiggins_appendix.html

Asking the Right Questions

Lynn Arthur Steen
*Department of Mathematics,
Statistics, and Computer Science
St. Olaf College*
steen@stolaf.edu

Assessment is about asking and answering questions. For students, "how am I doing?" is the focus of so-called "formative" assessment, while "what's my grade?" often seems to be the only goal of "summative" assessment. For faculty, "how's it going?" is the hallmark of within-course assessment using instruments such as ten-minute quizzes or one-minute responses on 3×5 cards at the end of each class period. Departments, administrations, trustees, and legislators typically ask questions about more aggregated levels: they want to know not about individual students but about courses, programs, departments, and entire institutions.

The conduct of an assessment depends importantly on who does the asking and who does the answering. Faculty are accustomed to setting the questions and assessing answers in a context where outcomes count for something. When assessments are set by someone other than faculty, skepticism and resistance often follow. And when tests are administered for purposes that don't "count," (for example, sampling to assess general education or to compare different programs), student effort declines and results lose credibility.

The assessment industry devotes considerable effort to addressing a variety of similar contextual complications, such as:

- different purposes (diagnostic, formative, summative, evaluative, self-assessment, ranking);
- different audiences (students, teachers, parents, administrators, legislators, voters);
- different units of analysis (individual, class, subject, department, college, university, state, nation);
- different types of tests (multiple choice, open ended, comprehension, performance-based, timed or untimed, calculator permitted, individual or group, seen or unseen, external, written or oral);
- different means of scoring (norm-referenced, criterion referenced, standards-based, curriculum-based);
- different components (quizzes, exams, homework, journals, projects, presentations, class participation);
- different standards of quality (consistency, validity, reliability, alignment);
- different styles of research (hypothesis-driven, ethnographic, comparative, double-blind, epidemiological).

Distinguishing among these variables provides psychometricians with several lifetimes' agenda of study and research. All the while, these complexities cloud the relation of answers to questions and weaken inferences drawn from resulting analyses.

These complications notwithstanding, questions are the foundation on which assessment rests. The assessment cycle begins with and returns to goals and objectives

(CUPM, 1995). Translating goals into operational questions is the most important step in achieving goals since without asking the right questions we will never know how we are doing.

Two Examples

In recent years two examples of this truism have been in the headlines. The more visible—because it affects more people—is the new federal education law known as No Child Left Behind (NCLB). This law seeks to ensure that every child is receiving a sound basic education. With this goal, it requires assessment data to be disaggregated into dozens of different ethnic and economic categories instead of typical analyses that report only single averages. NCLB changes the question that school districts need to answer from "What is your average score?" to "What are the averages of every subgroup?" Theoretically, to achieve its titular purpose, this law would require districts to monitor every child according to federal standards. The legislated requirement of multiple subgroups is a political and statistical compromise between theory and reality. But even that much has stirred up passionate debate in communities across the land.

A related issue that concerns higher education has been simmering in Congress as it considers reauthorizing the law that, among other things, authorizes federal grants and loans for postsecondary education. In the past, in exchange for these grants and loans, Congress asked colleges and universities only to demonstrate that they were exercising proper stewardship of these funds. Postsecondary institutions and their accrediting agencies provided this assurance through financial audits to ensure lack of fraud and by keeping default rates on student loans to an acceptably low level.

But now Congress is beginning to ask a different question. If we give you money to educate students, they say, can you show us that you really are educating your students? This is a new question for Congress to ask, although it is one that deans, presidents, and trustees should ask all the time. The complexities of assessment immediately jump to the foreground. How do you measure the educational outcomes of a college education? As important, what kinds of assessments would work effectively and fairly for all of the 6,600 very different kinds of postsecondary institutions in the United States, ranging from 200-student beautician schools to 40,000-student research universities? Indicators most often discussed include the rates at which students complete their degrees or the rates at which graduates secure professional licensing or certification. In sharp contrast, higher education mythology still embraces James Garfield's celebrated view of education as a student on one end of a log with Mark Hopkins on the other end. In today's climate of public accountability, colleges and universities need to "make peace" with citizens' demand for candor and openness anchored in data (Ekman, 2004).

I cite these examples to make two points. First, the ivory tower no longer shelters education from external demands for accountability. Whether faculty like it or not, the public is coming to expect of education the same kind of transparency that it is also beginning to demand of government and big business. Especially when public money is involved—as it is in virtually every educational institution—public questions will follow.

Second, questions posed by those outside academe are often different from those posed by educators, and often quite refreshing. After all these years in which school districts reported and compared test score averages, someone in power finally said "but what about the variance?" Are those at the bottom within striking distance of the average, or are they hopelessly behind with marks cancelled out by accelerated students at the top? And after all these years of collecting tuition and giving grades, someone in power has finally asked colleges and universities whether students are receiving the education they and the public paid for. Asking the right questions can be a powerful lever for change, and a real challenge to assessment.

Mathematics

One can argue that mathematics is the discipline most in need of being asked the right new questions. At least until very recently, in comparison with other school subjects mathematics has changed least in curriculum, pedagogy, and assessment. The core of the curriculum in grades 10–14 is a century-old enterprise centered on algebra and calculus, embroidered with some old geometry and new statistics. Recently, calculus passed through the gauntlet of reform and emerged only slightly refurbished. Algebra—at least that part known incongruously as "College Algebra"—is now in line for its turn at the reform carwash. Statistics is rapidly gaining a presence in the lineup of courses taught in grades 10–14, although geometry appears to have lost a bit of the curricular status that was provided by Euclid for over two millennia.

When confronted with the need to develop an assessment plan, mathematics departments generally take this traditional curriculum for granted and focus instead on how to help students through it. However, when they ask for advice from other departments, mathematicians are often confronted with rather different questions (Ganter & Barker, 2004):

- *Do students in introductory mathematics courses learn a balanced sample of important mathematical tools?*
- *Do these students gain the kind of experience in modeling and communication skills needed to succeed in other disciplines?*
- *Do they develop the kind of balance between computational skills and conceptual understanding appropriate for their long-term needs?*
- *Why can't more mathematics problems employ units and realistic measurements that reflect typical contexts?*

These kinds of questions from mathematics' client disciplines strongly suggest the need for multi-disciplinary participation in mathematics departments' assessment activities.

Similar issues arise in relation to pedagogy, although here the momentum of various "reform" movements of the last two decades (in using technology, in teaching calculus, in setting K–12 standards) has energized considerable change in mathematics instruction. Although lectures, problem sets, hour tests, and final exams remain the norm for mathematics teaching, innovations involving calculators, computer packages, group projects, journals, and various mentoring systems have enriched the repertoire of postsecondary mathematical pedagogy. Many assessment projects seek to compare these new methods with traditional approaches. But client disciplines and others in higher education press even further:

- *Do students learn to use mathematics in interdisciplinary or "real-world" settings?*
- *Are students encouraged (better still, required) to engage mathematics actively in ways other than through routine problem sets?*
- *Do mathematics courses leave students feeling empowered, informed, and responsible for using mathematics as a tool in their lives?* (Ramaley, 2003)

Prodded by persistent questions, mathematicians have begun to think afresh about content and pedagogy. In assessment however, mathematics still seems firmly anchored in hoary traditions. More than most disciplines, mathematics is defined by its problems and examinations, many with histories that are decades or even centuries old. National and international mathematical Olympiads, the William Lowell Putnam undergraduate exam, the Cambridge University mathematics Tripos, not to mention popular problems sections in most mathematics education periodicals attest to the importance of problems in defining the subject and identifying its star pupils. The correlation is far from perfect: not every great mathematician is a great problemist, and many avid problemists are only average mathematicians. Some, indeed, are amateurs for whom problem solving is their only link to a past school love. Nonetheless, for virtually everyone associated with mathe-

---

Greece, 250 BCE

If thou art diligent and wise, O stranger, compute the number of cattle of the Sun, who once upon a time grazed on the fields of the Thrinacian isle of Sicily, divided into four herds of different colours, one milk white, another a glossy black, a third yellow and the last dappled. In each herd were bulls, mighty in number according to these proportions: Understand, stranger, that the white bulls were equal to a half and a third of the black together with the whole of the yellow, while the black were equal to the fourth part of the dappled and a fifth, together with, once more, the whole of the yellow. Observe further that the remaining bulls, the dappled, were equal to a sixth part of the white and a seventh, together with all of the yellow. These were the proportions of the cows: The white were precisely equal to the third part and a fourth of the whole herd of the black; while the black were equal to the fourth part once more of the dappled and with it a fifth part, when all, including the bulls, went to pasture together. Now the dappled in four parts were equal in number to a fifth part and a sixth of the yellow herd. Finally the yellow were in number equal to a sixth part and a seventh of the white herd. If thou canst accurately tell, O stranger, the number of cattle of the Sun, giving separately the number of well-fed bulls and again the number of females according to each colour, thou wouldst not be called unskilled or ignorant of numbers, but not yet shalt thou be numbered among the wise.

But come, understand also all these conditions regarding the cattle of the Sun. When the white bulls mingled their number with the black, they stood firm, equal in depth and breadth, and the plains of Thrinacia, stretching far in all ways, were filled with their multitude. Again, when the yellow and the dappled bulls were gathered into one herd they stood in such a manner that their number, beginning from one, grew slowly greater till it completed a triangular figure, there being no bulls of other colours in their midst nor none of them lacking. If thou art able, O stranger, to find out all these things and gather them together in your mind, giving all the relations, thou shalt depart crowned with glory and knowing that thou hast been adjudged perfect in this species of wisdom.

—Archimedes. *Counting the Cattle of the Sun*

matics education, assessing mathematics means asking students to solve problems.

Mathematical Problems

Problems on mathematics exams have distinctive characteristics that are found nowhere else in life. They are stated with precision intended to ensure unambiguous interpretation. Many are about abstract mathematical objects—numbers, equations, geometric figures—with no external context. Others provide archetype contexts that are not only artificial in setting (e.g., rowing boats across rivers) but often fraudulent in data (invented numbers, fantasy equations). In comparison with problems people encounter in their work and daily lives, most problems offered in mathematics class, like shadows in Plato's allegorical cave, convey the illusion but not the substance of reality.

Little has changed over the decades or centuries. Problems just like those of today's texts (only harder) appear in manuscripts from ancient Greece, India, and China (see sidebars). In looking at undergraduate mathematics exams from 100 or 150 years ago, one finds few surprises. Older exams typically include more physics than do exams of today, since in earlier years these curricula were closely linked. Mathematics course exams from the turn of the twentieth century required greater virtuosity in accurate lengthy calculations. They were, after all, set for only 5% of the population, not the 50% of today. But the central substance of the mathematics tested and the distinctive rhetorical nature of problems are no different from typical problems found in today's textbooks and mainstream exams.

Questions suitable for a mathematics exam are designed to be unambiguous, to have just one correct answer (which may consist of multiple parts), and to avoid irrelevant distractions such as confusing units or complicated numbers. Canonical problems contain enough information and not an iota more than what is needed to determine a solution. Typical tests are time-constrained and include few problems that students have not seen before; most tests have a high proportion of template problems whose types students have repeatedly practiced. Mathematician and assessment expert Ken Houston of the University of Ulster notes that these types of mathematics tests are a "rite of passage" for students around the world, a rite, he adds, that is "never to be performed again" once students leave university. Unfortunately, Houston writes, "learning mathematics for the principal purpose of passing examinations often leads to surface learning, to memory learning alone, to learning that can only see small parts and not the whole of a subject, to learning wherein many of the skills and much of the knowledge required to be a working mathematician are overlooked" (Houston, 2001).

All of which suggests a real need to assess mathematics assessment. Some issues are institutional:

- *Do institutions include mathematical or quantitative proficiency among their educational goals?*
- *Do institutions assess the mathematical proficiency of all students, or only of mathematics students?*

Others are more specifically mathematical:

- *Can mathematics tests assess the kinds of mathematical skills that society needs and values?*
- *What kinds of problems would best reflect the mathematical needs of the average educated citizen?*
- *Can mathematics faculty fairly assess the practice of mathematics in other disciplines? Should they?*

China, 100 CE

- A good runner can go 100 paces while a poor runner covers 60 paces. The poor runner has covered a distance of 100 paces before the good runner sets off in pursuit. How many paces does it take the good runner before he catches up to the poor runner?

- A cistern is filled through five canals. Open the first canal and the cistern fills in 1/3 day; with the second, it fills in 1 day; with the third, in 2 1/2 days; with the fourth, in 3 days, and with the fifth in 5 days. If all the canals are opened, how long will it take to fill the cistern?

- There is a square town of unknown dimensions. There is a gate in the middle of each side. Twenty paces outside the North Gate is a tree. If one leaves the town by the South Gate, walks 14 paces due south, then walks due west for 1775 paces, the tree will just come into view. What are the dimensions of the town?

- There are two piles, one containing 9 gold coins and the other 11 silver coins. The two piles of coins weigh the same. One coin is taken from each pile and put into the other. It is now found that the pile of mainly gold coins weighs 13 units less than the pile of mainly silver coins. Find the weight of a silver coin and of a gold coin.

— *Nine Chapters on the Mathematical Art*

India, 400 CE

- One person possesses seven asava horses, another nine haya horses, and another ten camels. Each gives two animals, one to each of the others. They are then equally well off. Find the price of each animal and the total value of the animals possessed by each person.

- Two page-boys are attendants of a king. For their services one gets 13/6 dinaras a day and the other 3/2. The first owes the second 10 dinaras. Calculate and tell me when they have equal amounts.

— *The Bakhshali Manuscript*

Issues and Impediments

Assessment has had a tenuous impact in higher education, especially among mathematicians who are trained to demand rigorous inferences that are rarely attainable in educational assessment. Some mathematicians are unrelentingly critical of any educational research that does not closely approach medicine's gold standard of randomized, double blind, controlled, hypothesis-driven studies. Their fears are not unwarranted. For example, a recent federal project aimed at identifying high quality educational studies found that only one of 70 studies of middle school mathematics curricula met the highest standards for evidence (*What Works*, 2005). Virtually all assessment studies undertaken by mathematics departments fall far short of mathematically rigorous standards and are beset by problems such as confounding factors and attrition. Evidence drawn entirely from common observational studies can never do more than suggest an hypothesis worth testing through some more rigorous means.

Notwithstanding skepticism from mathematicians, many colleges have invested heavily in assessment; some have even made it a core campus philosophy. In some cases this special focus has led these institutions to enhanced reputations and improved financial circumstances. Nonetheless, evidence of the relation between formal assessment programs and quality education is hard to find. Lists of colleges that are known for their commitment to formal assessment programs and those in demand for the quality of their undergraduate education are virtually disjoint.

Institutions and states that attempt to assess their own standards rigorously often discover large gaps between rhetoric and reality. Both in secondary and postsecondary education, many students fail to achieve the rhetorical demands of high standards. But since it is not politically or emotionally desirable to brand so many students as failures, institutions find ways to undermine or evade evidence from the assessments. For example, a recent study shows that on average, high stakes secondary school exit exams are pegged at the 8th and 9th grade level to avoid excessive failure rates (Achieve, 2004). Higher education typically solves its parallel problem either by not assessing major goals or by doing so in a way that is not a requirement for graduation.

- *How, if at all, are the mathematical, logical, and quantitative aspects of an institution's general education goals assessed?*
- *How can the goals of comprehending and communicating mathematics be assessed?*

When mathematicians and test experts do work together to develop meaningful assessment instruments, they confront major intellectual and technical hurdles. First are issues about the harmony of educational and public purposes:

- *Can a student's mathematical proficiency be fairly measured along a single dimension?*
- *What good is served by mapping a multifaceted profile of strengths and weaknesses into a single score?*

Clearly there are such goods, but they must not be oversold. They include facilitating the allocation of scarce educational resources, enhancing the alignment of graduates with careers, and —with care—providing data required to properly manage educational programs. They do not (and thus should not) include firm determination of a student's future educational or career choices. To guard against misuse, we need always to ask and answer:

- *Who benefits from the assessment?*
- *Who are the stakeholders?*
- *Who, indeed, owns mathematics?*

Mathematical performance embraces many different cognitive activities that are entirely independent of content. If content such as algebra and calculus represents the nouns—the "things" of mathematics—cognitive activities are the verbs: know, calculate, investigate, invent, strategize, critique, reason, prove, communicate, apply, generalize. This varied landscape of performance expectations opens many questions about the purpose and potential of mathematics examinations. For example:

- *Should mathematics exams assess primarily students' ability to perform procedures they have practiced or their ability to solve problems they have not seen before?*

- *Can ability to use mathematics in diverse and novel situations be inferred from mastery of template procedures?*
- *If learned procedures dominate conceptual reasoning on tests, is it mathematics or memory that is really being assessed?*

Reliability and Validity

A widely recognized genius of American higher education is its diversity of institutions: students' goals vary, institutional purposes vary, and performance standards vary. Mathematics, on the other hand, is widely recognized as universal; more than any other subject, its content, practices, and standards are the same everywhere. This contrast between institutional diversity and discipline universality triggers a variety of conflicts regarding assessment of undergraduate mathematics.

Assessment of school mathematics is somewhat different from the postsecondary situation. Partly because K–12 education is such a big enterprise and partly because it involves many legal issues, major assessments of K–12 education are subject to many layers of technical and scholarly review. Items are reviewed for, among other things, accuracy, consistency, reliability, and (lack of) bias. Exams are reviewed for balance, validity, and alignment with prescribed syllabi or standards. Scores are reviewed to align with expert expectations and desirable psychometric criteria. The results of regular assessments are themselves assessed to see if they are confirmed by subsequent student performance. Even a brief examination of the research arms of major test producers such as ETS, ACT, or McGraw Hill reveal that extensive analyses go into preparation of educational tests.

In contrast, college mathematics assessments typically reflect instructors' beliefs about subject priorities more than any external benchmarks or standards of quality. This difference in methodological care between major K-12 assessments and those that students encounter in higher education cannot be justified on the grounds of differences in the "stakes" for students. Sponsors of the SAT and AP exams take great pains to ensure quality control in part because the consequences of mistakes on students' academic careers are so great. The consequences for college students of unjustified placement procedures or unreliable final course exams are just as great.

- *Are "do-it-yourself" assessment instruments robust and reliable?*
- *Can externally written ("off the shelf") assessment instruments align appropriately with an institution's distinctive goals?*
- *Can locally written exams that have not been subjected to rigorous reviews for validity, reliability, and alignment produce results that are valid, reliable, and aligned with goals?*

Professional test developers go to considerable and circuitous lengths to score exams in a way that achieves certain desirable results. For example, by using a method known as "item response theory" they can arrange the region of scores with largest dispersion to surround the passing (so-called "cut") score. This minimizes the chance of mistaken actions based on passing or failing at the expense of decreased reliability, say, of the difference between B+ and A– (or its numerical equivalent).

- *How are standards of performance—grades, cut-scores—set?*
- *Is the process of setting scores clear and transparent to the test-takers?*
- *Is it reliable and valid?*

Without the procedural checks and balances of the commercial sector, undergraduate mathematics assessment is rather more like the Wild West—a libertarian free-for-all with few rules and no established standards of accountability. In most institutions, faculty just make up tests based on a mixture of experience and hunch, administer them without any of the careful reviewing that is required for development of commercial tests, and grade them by simply adding and subtracting arbitrarily assigned points. These points translate into grades (for courses) or enrollments (for placement exams) by methods that can most charitably be described as highly subjective.

Questions just pour out from any thoughtful analysis of test construction. Some are about the value of individual items:

- *Can multiple choice questions truly assess mathematical performance ability or only some correlate? Does it matter?*
- *Can open response tasks be assessed with reliability sufficient for high-stakes tests?*
- *Can problems be ordered consistently by difficulty?*
- *Is faculty judgment of problem difficulty consistent with empirical evidence from student performance?*
- *What can be learned from easy problems that are missed by good students?*

Others are about the nature and balance of tests that are used in important assessments:

- *Is the sampling of content on an exam truly representative of curricular goals?*
- *Is an exam well balanced between narrow items that focus on a single procedure or concept and broad items*

that cut across domains of mathematics and require integrated thinking?

- *Does an assessment measure primarily what is most important to know and be able to do, or just what is easiest to test?*

Interpreting test results

Public interest in educational assessment focuses on numbers and scores—percent passing, percent proficient, percent graduating. Often dismissed by educators as an irrelevant "horse race," public numbers that profile educational accomplishment shape attitudes and, ultimately, financial support. K–12 is the major focus of public attention, but as we have noted, pressure to document the performance of higher education is rising rapidly.

Testing expert Gerald Bracey warns about common misinterpretations of test scores, misinterpretations to which politicians and members of the public are highly susceptible (Bracey, 2004). One arises in comparative studies of different programs. Not infrequently, results from classes of different size are averaged to make overall comparisons. In such cases, differences between approaches may be entirely artificial, being merely artifacts created by averaging classes of different sizes.

Comparisons are commonly made using the rank order of students on an assessment (for example, the proportion from a trial program who achieve a proficient level). However, if many students are bunched closely together, ranks can significantly magnify slight differences. Comparisons of this sort can truly make a mountain out of a molehill.

Another of Bracey's cautions is of primary importance for K–12 assessment, but worth noting here since higher education professionals play a big role in developing and assessing K–12 mathematics curricula. It is also a topic subject to frequent distortion in political contests. The issue is the interpretation of nationally normed tests that report percentages of students who read or calculate "at grade level." Since grade level is defined to be the median of the group used to norm the test, an average class (or school) will have half of its students functioning below grade level and half above. It follows that if 30% of a school's eighth grade students are below grade level on a state mathematics assessment, contrary to frequent newspaper innuendos, that may be a reason for cheer, not despair.

Bracey's observations extend readily to higher education as well as to other aspects of assessment. They point to yet more important questions:

- *To what degree should results of program assessments be made public?*

- *Is the reporting of results appropriate to the unit of analysis (student, course, department, college, state)?*
- *Are the consequences attached to different levels of performance appropriate to the significance of the assessment?*

Program Assessment

As assessment of student performance should align with course goals, so assessment of programs and departments should align with program goals. But just as mathematics' deep attachment to traditional problems and traditional tests often undermines effective assessment of contemporary performance goals, so departments' unwitting attachment to traditional curriculum goals may undermine the potential benefits of thorough, "gloves off" assessment. Asking "how can we improve what we have been doing?" is better than not asking at all, but all too often this typical question masks an assumed *status quo* for goals and objectives. Useful assessment needs to begin by asking questions about goals.

Many relevant questions can be inferred from *Curriculum Guide 2004,* a report prepared recently by MAA's Committee on the Undergraduate Program in Mathematics (CUPM, 2005). Some questions—the first and most important—are about students:

- *What are the aspirations of students enrolled in mathematics courses?*
- *Are the right students enrolled in mathematics, and in the appropriate courses?*
- *What is the profile of mathematical preparation of students in mathematics courses?*

Others are about placement, advising, and support:

- *Are students taking the best kind of mathematics to support their career goals?*
- *Are students who do not enroll in mathematics doing so for appropriate reasons?*

Still others are about curriculum:

- *Do program offerings reveal the breadth and interconnections of the mathematical sciences?*
- *Do introductory mathematics courses contain tools and concepts that are important for all students' intended majors?*
- *Can students who complete mathematics courses use what they have learned effectively in other subjects?*
- *Do students learn to comprehend mathematically-rich texts and to communicate clearly both in writing and orally?*

A consistent focus of this report and its companion "voices of partner disciplines" (mentioned above) is that the increased spread of mathematical methods to fields well beyond physics and engineering requires that mathematics departments promote interdisciplinary cooperation both for faculty and students. Mathematics is far from the only discipline that relies on mathematical thinking and logical reasoning.

- *How is mathematics used by other departments?*
- *Are students learning how to use mathematics in other subjects?*
- *Do students recognize similar mathematical concepts and methods in different contexts?*

Creating a Culture of Assessment

Rarely does one find faculty begging administrators to support assessment programs. For all the reasons cited above, and more, faculty generally believe in their own judgments more than in the results of external exams or structured assessments. So the process by which assessment takes root on campus is more often more top down than bottom up.

A culture of assessment appears to grow in stages (North Central Assoc., 2002). First is an articulated commitment involving an intention that is accepted by both administrators and faculty. This is followed by a period of mutual exploration by faculty, students, and administration. Only then can institutional support emerge conveying both resources (financial and human) and structural changes necessary to make assessment routine and automatic. Last should come change brought about by insights gleaned from the assessment. And then the cycle begins anew.

Faculty who become engaged in this process can readily interpret their work as part of what Ernest Boyer called the "scholarship of teaching," (Boyer, 1990) thereby avoiding the fate of what Lee Shulman recently described as "drive-by teachers" (Shulman, 2004). Soon they are asking some troubling questions:

- *Do goals for student learning take into account legitimate differences in educational objectives ?*
- *Do faculty take responsibility for the quality of students' learning?*
- *Is assessment being used for improvement or only for judgment?*

Notwithstanding numerous impediments, assessment is becoming a mainstream part of higher education programs, scholarship, and literature. In collegiate mathematics, however, assessment is still a minority culture beset by ignorance, prejudice, and the power of a dominant discipline backed by centuries of tradition. Posing good questions is an effective response, especially to mathematicians who pride themselves on their ability to solve problems. The key to convincing mathematicians that assessment is worthwhile is not to show that it has all the answers but that it is capable of asking the right questions.

References

Achieve, Inc. (2004). *Do Graduation Tests Measure Up?* Washington, DC.

"Assessment of Student Academic Achievement: Levels of Implementation." (2002). *Handbook of Accreditation.* Chicago, IL: North Central Association Higher Learning Commission.

Boyer, Ernest L. (1990). *Scholarship Reconsidered: Priorities of the Professoriate.* Princeton, NJ: Carnegie Foundation for the Advancement of Teaching.

Bracey, Gerald W. (2004). "Some Common Errors in Interpreting Test Scores." *NCTM News Bulletin*, April, p. 9.

Committee on the Undergraduate Program in Mathematics. (1995). "Assessment of Student Learning for Improving the Undergraduate Major in Mathematics," Focus, 15:3 (June) 24-28. www.maa.org/ saum/maanotes49/279.html

CUPM Curriculum Guide 2004. (2004). Washington, DC: Mathematical Association of America.

Ekman, Richard. (2004). "Fear of Data." *University Business*, May, p. 104.

Ganter, Susan and William Barker, eds. (2004). *Curriculum Foundations Project: Voices of the Partner Disciplines.* Washington, DC: Mathematical Association of America.

Houston, Ken. (2001). "Assessing Undergraduate Mathematics Students." In *The Teaching and Learning of Mathematics at University Level: An ICMI Study.* Derek Holton (Ed.). Dordrecht: Kluwer Academic, p. 407–422.

Ramaley, Judith. (2003). *Greater Expectations.* Washington, DC: Association of American Colleges and Universities.

Shulman, Lee S. (2004). "A Different Way to Think About Accountability: No Drive-by Teachers." *Carnegie Perspectives.* Palo Alto, CA: Carnegie Foundation for the Advancement of Learning, June.

What Works Clearinghouse. (2005). www.whatworks.ed.gov.

Assessing Assessment: The SAUM Evaluator's Perspective

Peter Ewell
National Center for Higher Education Management Systems (NCHEMS)
peter@nchems.org

The SAUM project took place within a broader context of assessment in American higher education. Faculty teams in mathematics departments experienced in microcosm what their colleagues in many other disciplines were simultaneously experiencing, and their actions were shaped by larger forces of politics and accountability affecting their institutions. At the same time, their efforts to develop and document viable department-level approaches to assessment in mathematics helped inform the national assessment movement—a field badly in need of concrete, discipline-level examples of good practice. Evaluation of SAUM helped bridge these two worlds.

In my personal role as project evaluator, I continued to participate in national conversations about assessment's purposes and prospects throughout the three-year grant period. But watching SAUM participants struggle with the day-to-day reality of crafting workable assessment approaches in their own departments helped keep me honest about what could and could not be accomplished. Similarly, the participant experiences that were revealed through the evaluation information we compiled often mirrored what was happening to other "early adopters" elsewhere.

The first section of this chapter sets the wider stage for SAUM by locating the project in a national context of assessment. A second section reflects on my role as project evaluator, and describes the kinds of evaluative information we collected to examine the project's activities and impact. A third section presents some of what we learned—focused primarily on what participants told us about how they experienced the project and the challenges they faced in implementing assessment initiatives back home.

SAUM in a National Context

The so-called "assessment movement" in higher education began in the mid-1980s with the confluence of two major forces. One originated inside the academy, prompted by growing concerns about curricular coherence and the conviction that concrete information about how and how well students were learning could be collectively used by faculty to improve teaching and learning (NIE, 1984). This version of "assessment" was low-stakes, incremental, faculty-owned, and guided by a metaphor of scholarship. The other driving force for assessment originated outside the academy prompted by policymakers' growing concerns about the productivity and effectiveness of colleges and universities (NGA, 1986). This version of "assessment" was high-stakes, publicly visible, accountability-oriented, and infused with the urgency of K–12 reform embodied in *A Nation at Risk* (USDOE, 1983).

Although fundamentally contradictory, both these forces were needed to launch and sustain a national movement. External authorities—first in the guise of states and later in the guise of regional accrediting organizations—served to constantly keep assessment at the forefront of institutional attention. But because these external requirements were at first fairly benign—and because academic leaders quickly saw the need to protect the academy's autonomy by developing locally-owned processes that might actually be useful—internal preferences for diverse evidence-gathering approaches aimed at institutional improvement served to discharge accountability as well for many years (Ewell, 1987).

The environment within which the SAUM project was launched was shaped by fifteen years of growing institutional experience with steering between these contradictory poles of assessment. By 2000, virtually every institution could claim that it "did assessment," at least in the sense that it had developed learning outcomes goals for general education and that it periodically surveyed its students and graduates. Most could also point to the beginnings of an institution-level organizational infrastructure for assessment—a coordinator operating out of the academic affairs office perhaps, or a faculty-staffed institutional assessment committee. About a third could lay claim to more sophisticated efforts including testing programs in general education, portfolios assembled by students and organized around general learning outcomes like "effective communication" or "critical thinking," or specially-designed assignments intended to both grade students individually and provide faculty with broader information about patterns of student strength and weakness in various abilities. Indeed, as revealed by the programs at such gatherings as the annual Assessment Forum hosted by the American Association for Higher Education (AAHE), there was a steady increase in the sophistication of institutional assessment efforts with respect to method and approach throughout this period, and equally steady progress in faculty acceptance of the fact that assessment was a part of what colleges and universities, for whatever reason, had to do.

By 2000, moreover, the primary reason why institutions had to "do assessment" had become regional accreditation. State mandates for assessment in public institutions, instituted in the wake of the National Governors Association's *Time for Results* report in the mid-1980s, had lost a lot of steam in the recession that appeared about 1990. States had other things to worry about and there were few resources to pursue existing mandates in any case. Accreditors, meanwhile, were under mounting pressure by federal authorities to increase their focus on student learning outcomes.

Regional accrediting organizations must be "recognized" by the U.S. Department of Education in order for accredited status to serve as a gatekeeper for receipt of federal funds. The federal recognition process involves a regular review of accreditation standards and practices against established guidelines. And since 1989 these guidelines have emphasized the assessment of learning outcomes more forcefully each time they have been revised by the Department. Accreditors are still accorded the leeway to allow institutions to develop their own learning outcomes and to assess them in their own ways. But by 2002, when SAUM was launched, it was apparent that accreditors could no longer afford to allow institutions to get by with little or no assessment—which had up to then essentially been the case—if they hoped to maintain their recognized status. The result was growing pressure on institutions to get moving on assessment, together with growing awareness among institutional academic leaders that a response was imperative.

But even at this late date, assessment remained something distant and faintly "administrative" for the vast majority of college faculty. It was rarely an activity departments engaged in regularly outside professional fields like engineering, education, business, or the health professions where specialized accreditation requirements made assessment mandatory. And even in these cases, the fact that deans and other academic administrators were front and center in the process, complete with the requisite guidelines, memos, schedules, and reports—all written in passive prose—made it likely that faculty in departments like mathematics would keep their distance. At the same time, despite their growing methodological prowess, few institutions were able to effectively "close the loop" by using assessment results in decision-making or to improve instruction. Periodic assessment reports were distributed, to be sure, but most of them ended up on shelves to be ritually retrieved when external visitors inquiring about the topic arrived on campus. Much of the reason for this phenomenon, in hindsight, is apparent. Assessment findings tend to be fine-grained and focused, while institutional decisions remain big and messy. *Real* application required smaller settings, located much closer to the teaching and learning situations that assessment could actually inform.

In this context, the notion of grounding assessment in the individual disciplines where faculty professionally lived and worked made a great deal of sense. For one thing, assessment practitioners had discovered that methods and approaches ought appropriately to vary substantially across fields and that such purportedly "generic" academic abilities as "critical thinking" and "communication" were manifest (and thus had to be assessed) very differently in different disciplinary contexts. At the disciplinary level, moreover, learning outcomes were generally much more easily specified than at the institutional level where of necessity they had to be so broadly cast that they often lost their meaning. More importantly, faculty tended to listen to one another more carefully in disciplinary communities bound by common languages and familiar hierarchies of respect. Even when assessment leaders on campus were faculty instead of administrators, their obvious background in methods derived from education and the social sciences often distanced them from colleagues in the sciences, humanities, fine and performing arts—as well as mathematics. For all these reasons, anchoring assessment in individual disciplinary communities was critical if it was to become a meaningful activity for faculty.

But why mathematics? In my view, mathematics became an "early adopter" of assessment for at least three reasons. First, the discipline is embedded in multiple aspects of teaching and learning beyond its own major at most institutions. Like colleagues in writing—but unlike those in physics, philosophy, and French—mathematics faculty had to staff basic skills courses in general education. As a result, both their course designs and pedagogies in such offerings as calculus and statistics must be closely aligned with a range of client disciplines including the sciences, engineering, business, and the social sciences. As a "basic skill," moreover, mathematics is generally assessed already at most institutions in the form of placement examinations, so at least some members of every mathematics department have experience with test construction and use. Where developmental mathematics courses are offered, moreover, they are often evaluated directly because the question of effectiveness is of broad institutional interest—a condition not enjoyed by, say, a course in Chaucer. All these factors meant that at least some members of any institution's mathematics department have at least some familiarity with broader issues of testing, evaluation, and pedagogy.

Second, mathematics has more explicit connections than most other disciplines with the preparation of elementary and secondary school teachers. Even if mathematics faculty are not explicitly located in mathematics education programs, many at smaller colleges and universities that produce large numbers of teachers are aware of pedagogical and assessment issues through this connection, and this knowledge and inclination can translate quickly to the postsecondary level.

Finally, at least at the undergraduate level, learning outcomes in mathematics are somewhat more easily specified than in many other disciplines. Although I have learned through the SAUM project that mathematicians are as apt to

disagree about the nuances of certain aspects of student performance as any other body of faculty—what constitutes "elegance," for instance, or an effective verbal representation of a mathematical concept—they can certainly come to closure faster than their colleagues elsewhere on a substantial portion of what undergraduate students ought to know and be able to do in the discipline. For all these reasons, mathematics was particularly well positioned as a discipline in 2002 to broaden and deepen conversations about assessment through a project like SAUM.

Evaluating SAUM: Some Reflections.

Serving as SAUM's external evaluator provided me with a personally unmatched opportunity to explicitly test my own beliefs and assumptions as an assessment practitioner. On the one hand, I have spent almost 25 years advocating for assessment, helping to develop assessment methods and policies, and working with individual campuses to design assessment programs. One cannot do this and remain sane unless one is at some level convinced of assessment's efficacy and benefit. Yet evaluation is an empirical and unforgiving exercise. SAUM's central premise was that it is possible to create a practice-based infrastructure for assessment that departments of mathematics could adapt and adopt for their own purposes, and thus improve teaching and learning. On the larger stage of institutional and public policy, this premise has been the basis for my professional career. The opportunity to "assess assessment" as it was acted out by one important discipline —and to reflect on what I found— was both exciting and sobering.

On a personal note, I also came to strongly value my role as "participant-observer" in the project and the opportunities that it provided me (and I hope to the project's participants) to see beyond customary professional boundaries. For my own part, I was gratified to witness many of the lessons about how to go about assessment that I had been preaching to Provosts and Deans for many years confirmed in microcosm among mathematics faculty at the departmental level. But I also saw (at times to my chagrin) the many differences in perception and failures of communication that can occur when such organizational boundaries are crossed.

As one telling example, at one of the SAUM department-level workshops I encountered a departmental team that reported a particularly frustrating bureaucratized approach to assessment at its institution being undertaken in response to an upcoming accreditation review—an institution that I knew from another source was being cited as a "model" of flexible and creative assessment implementation by the accreditor in question. I like to think that such insights, and they occurred throughout the project, helped keep me humble in the balance of my work in assessment.

At the same time, I like to think that my boundary-spanning role helped participants achieve some of the project's objectives. An instance here, as the previous example suggests, was my considerable ongoing work with accrediting organizations, which allowed me to interpret their motives and methods for SAUM participants, and perhaps set a broader context for their local assessment efforts.

Like many large, multi-faceted projects, SAUM presented many evaluation dilemmas. Certainly, it was perfectly straightforward to conduct formative data-collection efforts intended to guide the future implementation of project activities. For example, we collected participant reactions from the sixteen SAUM workshops conducted at MAA section meetings and used them to focus and improve these sessions. Responses from section meeting participants early in the process stressed the need for concrete examples from other mathematics departments that faculty could take home with them. Participating faculty observed that they often learned as much from interaction with other participants as from the material presented. These lessons were steadily incorporated into sessions at later section meetings (as well as into the design of the SAUM department-level workshops that were beginning to take shape at that time) and participant reactions steadily improved. Similarly, we learned through a follow-up of SAUM department-level workshop participants that a three-meeting format was superior to a two-meeting format and returned to the former in the project's last phase.

But determining SAUM's effectiveness in a more substantive way posed significant challenges. The most important of these was the fact that the bulk of the project's anticipated impact on mathematics faculty and departments would occur (if it did) well after the project was over. (As good an example of this dilemma as any is the fact that the publication in which this essay appears is one of the project's principal products; yet it reaches your hands as a reader only after the conclusion of the formal project evaluation!) Determining SAUM's effect on assessment practice in mathematics departments thus had to be largely a matter of following the experiences of project participants—particularly those mathematics faculty who attended the multi-session department-level workshops—when they returned to their home departments to apply what they learned. We did this primarily through email surveys given to participants ten months to a year after the conclusion of their workshop experience. The multi-session format of the

department-level workshops also helped the evaluation because at each of the workshop's concluding sessions we were able to explicitly ask participants about their experiences between workshop sessions. What we learned about the experience of assessment at the departmental level is reported in the following section.

We also set the stage for a more formal evaluation of SAUM's impact by conducting an electronic survey of mathematics departments early in the project's initial year. This was intended to provide baseline information about existing department-level assessment practices. A similar survey of departmental assessment practices will be undertaken at the conclusion of the project in the fall of 2005. The baseline survey was administered via MAA departmental liaisons to a sample of 200 mathematics departments stratified by size, institutional type, and location. 112 responses were received after three email reminders sent by the MAA, yielding a response rate of 56%. Questions on the electronic survey were similar to questions that we also posed to 316 individuals who attended SAUM workshop sessions at section meetings, which constituted another source of baseline information.

Justifying the project's potential impact, both sets of baseline data suggested that in 2002 most mathematics departments were at the initial stages of developing a systematic assessment approach. About 40% of department liaisons (and only 20% of participants at section meetings) reported comprehensive efforts in which assessment was done regularly in multiple areas, and another 35% (and 31% of participants at section meetings) reported that assessment was done "in a few areas." About 10% percent of department liaisons (and 21% of participants at section meetings) reported that assessment was "just getting started," and 15% percent (and more than a quarter of participants at section meetings) reported that there was "no systematic effort." Respondents from research universities reported somewhat lower levels of activity than other types of institutions. Differences in responses between department liaisons and the regular mathematics faculty who presumably attend section meetings are notable and reflect the pattern of reporting on institutional assessment activities typical of the late 1980s: in these surveys, administrators routinely reported higher levels of institutional engagement in assessment than was apparent to faculty at their own institutions (El-Khawas, 1987).

Baseline survey results also revealed that the mathematics major is the most popular target for assessment activities, with almost three quarters of responding departments indicating some activity here. About half of the departmental liaisons surveyed indicated that assessment takes place in either general service or remedial courses, and about a third reported that assessment takes place in courses for prospective teachers and in placement and advising. Not surprisingly, community colleges were somewhat more likely to report assessment in remedial and developmental courses, and less likely to report assessment of the major. Masters degree granting universities were more likely to be engaging in assessment of general service courses and courses for prospective teachers. Doctorate-granting research universities were somewhat less likely than others to be undertaking assessment in any of these areas.

Survey results suggested that mathematics departments are using a wide variety of assessment methods. The most popular method was faculty-designed examinations, which 62% of departments reported using. 53% of departments reported using standardized tests, which is not surprising, but more than 40% of departments reported employing so-called "authentic" approaches like work samples, project presentations (oral and written), or capstone courses. About 40% also reported using surveys of currently-enrolled students and program graduates. Standardized examinations tended to be used slightly more by departments engaging in assessment of remedial and developmental courses, while projects and work samples were somewhat more associated with assessing general service courses.

Finally, the departmental baseline survey asked respondents about their familiarity with *Assessment Practices in Undergraduate Mathematics* (Gold et al., 1999), which had then been in print for several years. Some 19% reported that they had consulted or used the volume, while another 35% noted that they were aware of it, but had not used it. The balance of 46% indicated that they were not aware of the volume. As might be expected, awareness and use were somewhat related to how far along a department felt it was with respect to assessment activity. About 61% of respondents from departments reporting comprehensive assessment programs in place said they were at least aware of *Assessment Practices* and about 25% had actively used it. Only about a third of those reporting no systematic plans had even heard of the volume and none had used it. Certainly, these baseline results leave plenty of room for growth and it will be instructive to see if three years of SAUM have helped move the numbers.

Emerging Impacts

Despite the fact that most of the SAUM project's impact will only be apparent after the publication of this volume, evaluation results to date suggest some emerging impacts. The majority must be inferred from responses to the follow-up surveys administered to department-level workshop participants about a year after they attended, focused on their continuing efforts to implement assessment projects in their own departments. Many of these results parallel what others have found in the assessment literature about the effective implementation of assessment at the institutional level, and for disciplines beyond mathematics.

Colleagueship. Like any change effort in the academy, implementing assessment can be a lonely business because its faculty practitioners are dispersed across many campuses with few local colleagues to turn to for practical advice or support. Indeed, one of the most important early accomplishments of the national assessment movement in higher education was to establish visible and viable networks of institution-level assessment colleagues through such mechanisms as the AAHE Assessment Forum (Ewell, 2002a).

Results of the evaluation to date indicate that SAUM is clearly fulfilling this role within the mathematics community. A first dimension here is simply the fact that SAUM is a network of *mathematicians*, not just "people doing assessment." As one faculty member told us, "history, agriculture, and even physics have different flavors of assessment from mathematics" and the opportunity to work with other mathematicians on mathematics topics in assessment was critical in grounding effective departmental efforts. Another dimension is simply the reassurance for individual mathematicians who first get involved in local projects that assessment is a going concern. Here it was useful for SAUM participants to learn that many mathematicians are already involved in assessment—more than many realized—and that assessment is not a peripheral activity that only a few mathematics departments are involved in. These points were seen by SAUM participants as particularly important in "selling" assessment to other faculty when they returned to their home departments.

The team basis for participation in SAUM workshops meant automatic colleagueship and mutual support. Simply being away together with colleagues, far from the pressures of everyday campus work was also deemed helpful. At the same time, working with other campuses at the workshop in multiple encounters helped build a feeling among SAUM participants of being part of a larger "movement" that had momentum. This was especially important for faculty who felt, in the words of one, that they had "been thrust into a leadership position on assessment" with little real preparation for this role. Knowing that others were in the same position and sharing approaches about what to do about it was seen as especially important.

The same was true of learning about more specific assessment approaches where, as expected, SAUM departments borrowed liberally from one another. But by far the most important impact of having colleagues was the stimulus they provided to keep participating departments moving. The need to present departmental progress periodically and publicly was important to this dynamic: teams at the workshops knew that they were going to have to report to their peers, so worked hard to have something to say. As one representative noted, "if we had run out of time and didn't accomplish what we intended, it probably wouldn't have had any consequences on campus—we were already doing more than most departments—but because we had to have presentations ready at different workshops we were pushed to follow through on plans and to discuss and revise our activities." This "peer stimulation" effect was a particularly important dynamic in SAUM, and parallels similar lessons learned about colleagueship in other assessment-related change projects (e.g., Schapiro and Levine, 1999).

On a more sobering note, however, early evaluation results also suggest the difficulty of maintaining colleagueship absent the explicit framework of a project or a visible network to support it. Few departmental representatives reported contacting other participants on their own, largely due to pressures of time. Again confirming lessons of the assessment movement more generally, an infrastructure for sustaining assessment in mathematics must be actively *built*; it will not just happen as a result of peoples' good experiences.

Learning By Doing. Another lesson of the assessment movement nationally was the importance of early hands-on experience and practice. Evaluation results to date suggest that SAUM is strongly replicating this finding within mathematics departments. Rather than looking for the best "model program" and planning implementation down to the last detail, SAUM departments, like their colleagues in many disciplines, learned quickly that time invested in even a messy first effort trumped similar investments in "perfecting" design. As one participant told us, "the most important lesson that I learned was to just get started doing something." Another said, "begin with a manageable project, but begin." Small projects can not only illustrate the assessment process in manageable ways with only limited investments of resources but can also quickly provide tangible accomplishments to show doubting colleagues. The importance of this insight was reinforced by the fact that most participants also discovered that assessment was a good deal more time-consuming than they had first imagined, even for relatively simple things.

At the same time, participating departments learned the importance of finding their own way *in* their own way, and that local variations in approach are both legitimate and

effective. Like their colleagues elsewhere in assessment, they also learned tactical lessons about implementation that could only be learned by doing. As one department reported, "for us, designing an assessment program means finding a balance between getting good information ... and not increasing faculty workload too much." Additional comments stressed the importance of knowing that "one size does not fit all" and that good assessment should be related to local circumstances.

Finally, participants encountered aspects of local departmental culture that could not be addressed through formulaic methods. One of them summarized this condition nicely: "there are rules at my institution about how we have to do assessment even though those rules are unwritten, unarticulated (except when violated), and specific to my institution and the larger community. I used to think that these rules were to be found somewhere in the literature ... now I know that I'm dealing with the unknown and with rules that are likely being made up as we go. This makes me much more confident in my own ideas instead of backing down when I am told that something is 'not allowed.'"

Growing Maturity. As their projects evolved, most participating departments reported a growing maturity with assessment. Several departments doing program-level assessment, for example, had replicated their assessment models in another related department or program (e.g., computer science), or had been working with other departments at the institution to help them develop an assessment approach. Others implemented or regularized activities that they had planned or experimented with at the workshop. Most indicated that they had expanded their departmental assessment efforts to become more systematic and comprehensive—adding new assessment techniques and applying them to more courses or involving more faculty. As one participating department reported, "We believe one of our greatest accomplishments is to have engaged a significant proportion of the department (more than half the faculty) in assessment in one way or another."

Growing maturity is also apparent in organizational and motivational dimensions. With regard to the former, several departments reported that they had discovered the importance of having a departmental advocate or champion for assessment who could set timelines, enforce deadlines, and provide visibility. A few also reported "regularizing" assessment activities—in one case allowing the original project leader to hand off assessment activities to a newly interested faculty member to coordinate or lead. As one departmental representative put it, participating "got us off to a great start and developed a sense of confidence that we are in a better posture with assessment than other departments on our campus."

Motivational shifts were more subtle, but reflected a shift toward internal instead of external reasons for engaging in assessment. Mirroring experience in other fields, many mathematicians first heard about assessment through accreditation or their administration's desire to "create a program." But as their participation in SAUM progressed, many also reported new attitudes toward assessment. As one faculty member put it, "Earlier [activities] were about responding to outside pressure…[later activities] were about doing this for ourselves." Another noted, "Most people [at my institution] are not as advanced in assessment as we are in the mathematics department. ... The task still seems to most people like a necessary activity conducted for external reasons, rather than an activity that has intrinsic value to improve their own work." This shift requires time to accomplish and findings from other fields emphasizes the fact that outside pressures or occasions are important to start things moving on assessment at the institutional level (Ewell, 2002b). But SAUM participants began to recognize also that sustaining assessment requires the kind of internal motivation that can only be developed over time and through collective action.

Changing Departmental Culture. Twenty years after the emergence of assessment as a recognizable phenomenon in higher education, it has yet to become a "culture of use" among faculty in disciplines that lack professional accreditation. Many reasons for this have been advanced, ranging from alien language to lack of institutional incentives for engagement, but by far the most prominent is the imposition of assessment requirements by external authorities (Ewell 2002a). Consistent with national experience in other disciplines, SAUM workshop participants thus returned to their own departments determined to make a difference, but they faced an uphill battle to change their colleagues' attitudes about assessment and, in the longer term, to begin to transform their department's culture.

A first milestone here was the fact that participation in SAUM itself helped legitimize the work of developing assessment. Being part of a recognized, NSF-funded project was important in convincing others that the work was important. So was the clear commitment of workshop participants to working on their projects. As one faculty member told us, "Because [the participants] were genuinely interested, ... that interest and enthusiasm has been acknowledged by others." Several also mentioned the value of knowing the "justifications for assessment" in communicating with fellow faculty members.

But SAUM participants also tended to end up being the "assessment people" in their departments—accorded legitimacy for their activities to be sure, but not yet joined by sig-nificant numbers of colleagues. As one wryly stated, "it has probably made more work for me as when I share an idea of something that we can do, I usually get put in charge of doing it." Another doubted that he and his SAUM teammate had gained much stature in the department because of their participation, "but I guess at least more people recognize what we have done."

As national experience suggests, moreover, wider impacts on departmental culture with respect to assessment require time to develop—more than the three years of engagement most SAUM participants have to date enjoyed. Most indicated that their colleagues were in general more informed about assessment as a result of SAUM and were therefore more willing to agree that it might be beneficial for their departments or institutions. So despite little groundswell of enthusiasm, most did report slow progress in changing departmental attitudes. One participant captured the typical condition succinctly when he reported that his colleagues "remain largely indifferent to assessment…they are in favor of improving programs as long as it doesn't bother them." Another described this condition as follows: "I think there is still a degree of skepticism about all of this, but at least we don't run into outright hostility or claims that this is all a great waste of time and effort." Echoing these comments, a third reported that "the department has definitely become more open to the idea of assessment ... for one thing, they have finally realized that it is not going away ... for another, if there is someone willing to do the work, they will cooperate." These are no small achievements. But the overall pattern of impact to this point remains one of increased awareness and momentum for assessment among SAUM departments with only a few early signs of a changed departmental culture.

"Assessing assessment" through the SAUM evaluation remains an ongoing activity. Like assessment itself in many ways, the task will never be finished. But it is safe to conclude at this point that mathematics has built a resource through SAUM that if maintained, will be of lasting value. On a personal note, I have learned much from my colleagues in mathematics and have been grateful for the opportunity to work with them on a sustained basis. And from a national perspective, I can say without reservation that they are making a difference.

References

El-Khawas, Elaine (1987). *Campus Trends: Higher Education Panel Report #77.* Washington, DC: American Council on Education (ACE).

Ewell, Peter T. (1987). *Assessment, Accountability, and Improvement: Managing the Contradiction.* Washington, DC: American Association of Higher Education (AAHE).

Ewell, Peter T. (2002a). An Emerging Scholarship: A Brief History of Assessment. In T.W. Banta and Associates, *Building a Scholarship of Assessment*. San Francisco: Jossey Bass, 3–25.

Ewell, Peter T. (2002b). A Delicate Balance: The Role of Education in Management. *Quality in Higher Education*, 8, 2, 159–172.

Gold, Bonnie, Sandra Z. Keith, and William A. Marion, eds. (1999) *Assessment Practices in Undergraduate Mathematics*. MAA Notes, Vol. 49. Washington, DC: Mathematical Association of America. www.maa.org/saum/maanotes49/index.html.

National Governors Association (1986). *Time for Results: The Governors' 1991 Report on Education*. Washington, DC: National Governors Association.

National Institute of Education, Study Group on the Conditions of Excellence in American Higher Education (1984). *Involvement in Learning: Realizing the Potential of American Higher Education*. Washington, DC: U.S. Government Printing Office.

Schapiro, Nancy S. and Levine, Jodi H. (1999). *Creating Learning Communities: A Practical Guide for Winning Support, Organizing for Change, and Implementing Programs*. San Francisco: Jossey-Bass.

U. S. Department of Education, National Commission on Excellence in Education (1983). *A Nation at Risk: The Imperative for Educational Reform*. Washington, DC: U.S. Government Printing Office.

Developmental, Quantitative Literacy, and Precalculus Programs

Assessment of Developmental, Quantitative Literacy, and Precalculus Programs

Bonnie Gold
Department of Mathematics
Monmouth University
West Long Branch, NJ
bgold@monmouth.edu

From an assessment perspective, developmental, quantitative literacy, and pre-calculus courses have many similarities and interrelationships. At many institutions, these courses constitute most of the department's workload. They are not generally the courses in which most faculty members invest their greatest enthusiasm or concern: that is usually reserved for courses for mathematics majors (or perhaps, in universities with graduate programs, for graduate students). They are the least mathematically interesting courses we teach. Moreover, since they're usually filled with students who dislike and fear mathematics and would rather be anywhere except in mathematics class, these are often the most difficult and frustrating courses to teach.

As a result, however, these are courses in which effective assessment can yield the greatest improvement in faculty working conditions as well as in student learning. If mathematics departments can turn these courses from ones students just muddle through into courses in which they grow in mathematical confidence and competence, these courses can become enjoyable and interesting to teach.

Colleges and universities are under pressure to develop assessment programs primarily of two types: for majors (often including other subjects required by a major), and for general education. The latter emphasis often leads to requests to mathematics departments to assess quantitative literacy. Pressures to assess developmental mathematics programs, on the other hand, typically reflect concerns about finances or about rates of student progress toward graduation. If students must repeat developmental courses several times before succeeding, or if they pass the developmental courses only to fail the credit-bearing courses for which they are prerequisites, students either graduate late or drop out entirely.

Interactions among these programs

There are no sharp boundaries between developmental, quantitative literacy, and precalculus courses. Some institutions require no mathematics; at some others, general education requirements are met by simply passing a placement exam or the developmental courses. Still others assume that adequate quantitative literacy skills will be developed through college algebra or precalculus courses. Often precalculus courses are directed at students planning to take calculus, but in fact are taken primarily to satisfy general education requirements. When this happens, faculty view the high DWF rate (D/Withdraw/Failure) in such courses as "casualties on the road to calculus." In reality, few of these students were ever on that road.

Programs assessing one of these components are likely to involve the others simply because of their interactions. The success of a developmental course is measured in large part by its students' success in their later quantitative literacy, college algebra, and precalculus courses. An assessment of precalculus courses needs to consider to what extent the courses are being used to meet needs for quantitative literacy, either by design or *de facto*, as well as how well they prepare students for calculus.

Since all of these courses are primarily for non-majors, good planning and good assessment should extend beyond the mathematics department. As with courses for students in mathematics-intensive majors, ensuring that the needs of other majors are met, as well as those of the institution, can do wonders for the reputation of the mathematics department on campus and can lead to support for its other initiatives (Chipman 1999).

The role of an effective placement process. In all of these courses, effective placement can be crucial to student success. It is very difficult to teach a class well when half the class has the prerequisites for the course and half does not. Many institutions, for example Arizona Western College (p. 47) and the University of Arizona (Krawczyk & Toubassi 1999), have found that replacing a voluntary placement process (where students may register for courses other than those they are placed into) by one in which students cannot take a course until they meet the prerequisites by placement or by taking courses makes a large difference in student success rates. This has also been my experience at Monmouth University.

Good placement processes generally involve multiple components (Cederberg 1999). A single examination cannot place students as accurately as can a process that combines this test result with information on high school rank, grade point average, last mathematics course taken, how long since that last course was taken, and SAT or ACT scores. As a project at Virginia Tech revealed, student self-descriptions in terms of how good they are at mathematics can also be an effective component of the placement procedure (Olin & Scruggs 1999).

Assessing the effectiveness of the placement process itself. Examining how well each factor and the overall formula predicts success and adjusting the formula in response to this analysis is an important part of the assessment of these introductory courses. National placement examinations such as Accuplacer[1] are appropriate only if the skills they test are those students need for the courses they are

actually taking. For example, the skills needed for success in quantitative literacy courses are generally quite different from those needed for college algebra or calculus. Many locally-written placement tests also ignore this criterion.

The assessment cycle, applied to these clusters of courses

Examining goals and learning objectives. More than with most other courses that a mathematics department offers, there is often a substantial gap between course content and the course's role in the curriculum. Ideally, a discussion of goals will lead to changes in curriculum (often new courses) and improved student learning. When development of course goals begins with questions about desired student outcomes (e.g., "what do we want students to get out of this course? what should they be able to do when they've finished?"), it heads off the litany of faculty complaints about the students' lack of abilities or work ethic. The resulting goals will include statements of mathematical skills (e.g., "students should recognize when a linear model is appropriate to consider, be able to develop this model from given data, and make predictions from the model"), but usually also some broader skills (e.g., "students should be able to read critically a newspaper article involving graphs") and perhaps some affective outcomes (e.g., "students should feel less mathophobic at the end of the course"). Partner disciplines and committees on general education can provide useful input as mathematics departments define their course goals. Mount Mary College (p. 59) and Allegheny College (p. 37) discovered that simply detailing course goals for their developmental and quantitative literacy courses led to development of more appropriate courses.

From goals to objectives. To be able to assess goals effectively, they need to be made concrete. This is done by developing, for each goal, one or more learning objectives specifying skills students must develop to meet that goal. Concrete learning objectives can be developed even for affective goals. For example, a learning objective for the goal "students should feel less mathophobic ..." might be that "students will attempt to solve problems of types never before encountered, rather than skipping them entirely." Of course, it is easier to develop learning objectives for goals that are more specifically related to mathematical content, and there are likely to be more learning objectives for each of these goals. A report from King's College offers helpful discussion and examples of the difference between goals and learning objectives (Michael, 1999).

Sharing goals and objectives with all constituents. In larger institutions there are typically several sections of these

---

[1] www.collegeboard.com/highered/apr/accu/accu.html

introductory courses each semester, often taught by adjunct faculty or graduate students. In such cases, greater uniformity of student learning can be achieved by sharing with all faculty involved in teaching these courses—and perhaps also with the students—a clear statement of the goals and learning objectives of the courses, of how each of the objectives is to be achieved in the course, and how it will be assessed. Oakland University developed helpful information sheets for just this kind of purpose (Chipman 1999).

Choosing appropriate assessment mechanisms. Carefully chosen learning objectives often lead naturally and easily to appropriate assessment mechanisms. Timed, in-class tests are appropriate for assessing the particular mathematical content required in successor courses. However, other tools are often more effective for assessing affective or conceptual skill development. For example, if your objective is that students at least attempt to solve a problem, using a test question for which partial credit will be given is more likely to give this information, as are activities done under less time pressure and in less stressful situations than in-class examinations.

Sometimes it is effective to have mathematics' partner disciplines administer some of the assessments. For example, at the beginning of a psychology class that is going to use students' quantitative or algebraic skills, a brief quiz over prerequisite concepts can give both the psychology instructor and the mathematics department useful information on what has been retained. See the case study from Portland State University in this volume (p. 65) and in an earlier report from the University of Wisconsin and North Dakota State University (Martin & Bauman 1999) for good examples of using client departments to give this kind of feedback.

The studies at San Jose State University (p. 75)and Virginia Polytechnic Institute (Olin & Scruggs 1999) show that students' attitudes toward mathematics and how it is learned, and toward the courses themselves, can significantly affect their performance. While *surveys of student attitudes* cannot *alone* assess student learning, giving such surveys early in a course and working on improving students' beliefs about what they must do to succeed in mathematics can affect students' success. At Richard Stockton College of New Jersey (Ellen Clay 1999) and at Ball State University (Emert 1999) students write a class mission statement for quantitative literacy courses, while at St. Cloud State University students are asked to reflect on what they can do to improve their chances for success (Keith 1999). Students at St. Cloud State use journal articles in their planned major to explore the relevance of their study of mathematics to their future careers.

Developmental, quantitative literacy, and precalculus courses are particularly good places to use a range of *formative assessment techniques*. These are ways to find out what students do and don't understand about what has been presented, and at the same time to help students develop desired skills. For example, by having students write explanations of their answers to questions, the instructor learns what their confusions are, and the students have to think through what they understand. An example from the University of Southern Colorado (Barnett 1999) uses expanded true-false questions: "Determine if each of the following is true or false, and give a complete written argument for your response." An instructor at Surry Community College used "concept maps" to learn what students think they know about a topic (Atkins 1999). Many programs have students learn by writing about mathematics and explore ideas by working in groups. A lot of information on using this kind of formative assessment can be found in the section on "Assessment in the Individual Classroom" of *Assessment Practices in Undergraduate Mathematics* (Gold, *et al. 1999*).

Completing the cycle: using the data. In any type of assessment, the point of the exercise is to complete the cycle by using the data collected to improve learning. Usually the results of the first assessment activity raise more questions than they answer. If students aren't doing well in a follow-up course, what is the cause? Looking at the program, you can come up with some conjectures; these lead to further assessment activities. Once you find the causes, it's time to look at how to *change the program*. Since almost all of the programs considered here involve partner disciplines or university committees, revision should include not only the mathematics department but these other constituencies as well. This takes more time but yields many benefits for the department due to the good will gained. Finally, of course, you need to start the cycle again: as you redesign courses, consider goals, rethink learning objectives, and decide how to assess the effectiveness of the changes.

Developmental courses

The goal for developmental courses seems clear: prepare students for credit-bearing courses. However, judging by the course content, the goal often *appears* to be remediating what students haven't learned in grades K–12. These two goals may be quite different. High schools attempt to prepare children for all possible educational futures, including majoring in mathematics. Once the student is in college, that educational aim may be much better defined, and the student may not ever need to study calculus. So remediating the high school deficiencies may not be necesssary or appropriate. Sometimes, however, the institution's view may be that every student should master certain mathematical skills. If some of these are normally mastered at the pre-college level, achieving these skills does become one of the goals of the developmental program. Depending on the college's view of its general education expectations, the developmental courses may be used to meet general education needs as well (in which case, as goals are developed, that expectation must be included). It is also not uncommon for developmental courses to be the prerequisites for various courses in the sciences and social sciences, in which case these partner disciplines should be invited to communicate their expectations.

Tracking student success rates. Assuming that the primary purpose of developmental courses is to prepare students for further courses in mathematics, the most direct way to assess this objective is to investigate student success in these later courses. For an overall, thumbs-up/thumbs-down, answer, this involves tracking these students in these later courses. Several case studies report on this kind of effort: Allegheny College (p. 37); Cloud County Community College (p. 55); and earlier, St. Peter's College (Poiani, 1999).

There are two obvious approaches to this investigation. One option is to get a list of all students in the follow-on courses and their grades, look to see which of these students took developmental courses, and compare that group's grades with students overall. The other approach is to look in the other direction: follow students from the developmental courses to see how they do in their later mathematics courses.

If you find as did Allegheny College (p. 37) that students who ignore placement into developmental courses and enroll directly in the credit-bearing courses do better than students who take and pass the developmental courses first, you clearly need to investigate further. It may be the placement process that needs revision, not the developmental offerings. Perhaps a factor such as student self-reporting on how hard they work or their mathematical confidence needs to be included. Or the developmental students may wait a long time before taking the credit-bearing course. You can control for this by looking at the time elapsed between taking the developmental course and its successor. But it also may be that the developmental courses really need serious restructuring.

Generally in assessment, the answer to one question leads to another question, or to making the question more precise. Effectiveness of a developmental program can be further investigated by giving pre-tests at the beginning of the credit-bearing courses to determine whether students, by the time they take the successor course, have the skills needed for success in that course. You may need to add some questions beyond computational multiple-choice problems: problem-solving questions, open-ended questions where you look at the methods students use, for example. This is a finer sieve than simply tracking student success rates, as you can determine precisely which skills students are either not learning or not retaining. This information can then be translated directly into changes of emphasis or teaching methods in the developmental courses, and perhaps (when it's an issue of non-retention due to a large time lapse between the courses) changes in student advising. (The report from Arizona Western College (p. 47) offers a good example of this.)

Attitude surveys. A second (and generally secondary) method sometimes used for assessing developmental courses is attitude surveys. Often these are used more for formative than summative assessment, but learning that students' attitudes about their mathematical abilities or the value of mathematics has not changed can be a red flag warning that the course isn't achieving its goals. The case study from Mount Mary College (p. 59) discusses such a survey. Surveys of faculty in subjects that have the developmental courses as their mathematical prerequisites, asking what mathematical weaknesses they feel their students have, can also indicate areas which need improvement.

Formative assessment. Developmental students are perhaps the group that benefits most from the use of a variety of formative assessment methods such as using group work or brief writing assignments during class time. These make the course less dry, force students to reflect on the computations they're learning, and take the course beyond simply repeating unsuccessful high school experiences at a faster pace.

Quantitative literacy courses

Quantitative literacy courses are mathematics' most amorphous courses. There is an enormous variety of mathematics that can inform the thinking of an educated citizenry, but this cannot all be crammed into the one or two courses required of students to meet general education requirements.

In 1996, the Quantitative Literacy Subcommittee of the Committee on the Undergraduate Program in Mathematics issued guidelines for quantitative literacy programs (Sons, 1996). An excellent description of quantitative literacy and a summary of the CUPM recommendations appeared in *Assessment Practices in Undergraduate Mathematics* (Sons, 1999). These reports argue that a college graduate should be able to:

- interpret mathematical models such as formulas, graphs, tables, and schematics, and draw inferences from them;
- represent mathematical information symbolically, visually, numerically, and verbally;
- use arithmetical, algebraic, geometric and statistical methods to solve problems;
- estimate and check answers to mathematical problems in order to determine reasonableness, identify alternatives, and select optimal results;
- recognize that mathematical and statistical methods have limits.

Other expressions of goals for quantitative literacy (Dwyer, to appear; Steen 1997; Steen 2001) convey different ranges of expectations for example:

- estimating answers and checking answers for reasonableness;
- understanding the meaning of numbers;
- using common sense in employing numbers as a measure of things.

The primary recommendation of the CUPM report is that quantitative literacy cannot be achieved in a single course, but should involve at least two courses, a foundational experience (often, but not necessarily, taught in the mathematics department) followed by a continuation experience, often within the student's major. An alternative to the two-course model for building a rich quantitative literacy experience at the college level is to infuse quantitative literacy across the general education program. It is important to get faculty members from a full range of disciplines involved in the development of courses to meet this recommendation, and a survey of faculty on what they see as the quantitative literacy needs of their students can be a good starting point.

Recognition that we cannot do everything in one course gives us the freedom to look at the set of skills and understandings we want students to develop in this kind of course, and find multiple ways of meeting these goals. This often leads to a great variety of content in these courses, even within the same institution. To assess how well these courses meet students' quantitative literacy needs, we must look not only at students' success in learning the mathematical content of a given course.

A well-designed examination may well be a good beginning, as it can be used as a pre- and post-test across a range of quantitative literacy courses, and even in the continuation experience courses. If course instructors are given the results of pretests, they can find a level of course presentation more appropriate to the abilities the students bring to the course. King's College (Michael 1999) and Virginia Commonwealth University (pp. XX–YY) have used some

kind of pre/post test for such courses; the Portland State psychology department (pp. XX–YY) offers an example of such testing in "continuation experience" courses.

However, many quantitative literacy goals cannot be assessed solely via a multiple choice examination. There should be at least a free-response portion to assess students' ability to represent mathematical information and to test their ability to interpret the results of calculations. A common pitfall is to write the test questions without correlating them with the program goals. It is also difficult to write questions that are not course-content specific and thus can be used to compare students' development across a range of such courses. Sons' article (1999) discusses issues such as being able to compare student learning across courses that may have very different content.

Many alternative assessment methods are particularly useful in quantitative literacy courses, both formatively and summatively: portfolios, journals, using writing to learn mathematics, having students create problems for use either as review for an exam or as exam questions themselves, projects, group work (Gold, et al., 1999, especially part II). To use these alternative methods for summative assessment of the quantitative literacy program, rubrics must be developed to enable readers to summarize rapidly masses of information in ways that give useful information that can be compared across courses. Sometimes this can be done as faculty members grade the activity in the course (but separately from the grade for the activity) if the rubrics have been developed and faculty members trained in their use in advance.

College algebra and precalculus courses

When most go on to calculus. The biggest assessment challenge in college algebra and precalculus is deciding what the goals are. As long as the courses are primarily being used to develop skills necessary for students to succeed in calculus, the goals are fairly clear, and the assessment issues are similar to those for developmental courses. It is, however, important to determine whether this is in fact their primary use. If it is, initial assessment methods can involve looking at how well students succeed in the next course, viz. calculus. The case study from San Jose State University (p. XX–YY) illustrates just this situation.

A more detailed analysis requires study of the particular skills students need in calculus and how well they do on those skills, both on final exam questions in the precalculus course and on similar questions given at the beginning of the calculus course. Often items from the department's placement test can be the initial questions used in such an assessment. It is important, however, to correlate students' scores on these questions at the beginning of calculus with their grades at the end to make sure that these questions really do test skills necessary for success in calculus.

However, multiple choice placement test items cannot test all the skills students will need for success in calculus. Students need to be able to translate word problems into algebraic equations, to translate between multiple representations of data and functions, and much more. Often examining student work on one long-answer question can pinpoint the range of confusions that are resulting in incorrect answers on multiple-choice questions.

Surveys can be used in these courses, as in developmental courses, to examine student attitudes towards mathematics and how it is learned. Often students who had calculus in high school place into precalculus in college not because of lack of ability but because they haven't yet learned how to study mathematics. At San Jose State (p. XXX-YYY) successful students showed a significant understanding of what was needed to succeed in these courses.

When few go on to the calculus sequence. On the other hand, if you find as did Allegheny College (p. XX–YY) that the majority of students in college algebra/precalculus are not going on to a full year of calculus, you may want to consider restructuring the courses. Often courses called college algebra or precalculus are used to meet at least three different needs: they serve as the quantitative literacy (and thus, terminal mathematics) course for a large number of students; they prepare students in business and biology for an applied calculus course; and they also try to prepare students to succeed in the mainstream calculus sequence. They are trying to meet the needs of three very different audiences with extremely different goals. I've yet to see a program that recognizes and acknowledges all these goals and successfully meets them in a single course. Discussions with faculty in partner disciplines can help mathematics departments decide what combination of courses will be most effective in their particular context. If after such discussions the pre-calculus or college algebra course is still left meeting the needs of partner disciplines (in addition to preparing students for mainstream calculus) or helping students develop quantitative reasoning skills, the assessment program for these courses must include activities that assess these other goals in addition to the activities mentioned here that assess preparation for calculus.

References

Atkins, Dwight. "Concept Maps." In *Assessment Practices in Undergraduate Mathematics*, Bonnie Gold, *et al.*, eds. Washington DC, Mathematical Association of America, 1999, pp. 89–90.

Barnett, Janet Heine. "True or False? Explain!" In *Assessment Practices in Undergraduate Mathematics*, Bonnie Gold, et al., eds. Washington DC, Mathematical Association of America, 1999, pp. 101–103.

Cederberg, Judith N., "Administering a Placement Test: St. Olaf College." In *Assessment Practices in Undergraduate Mathematics*, Bonnie Gold, et al., eds. Washington DC, Mathematical Association of America, 1999, pp. 178–180.

Chipman, J. Curtis. "Let Them Know What You're Up to, Listen to What They Say." In *Assessment Practices in Undergraduate Mathematics*, Bonnie Gold, et al., eds. Washington DC, Mathematical Association of America, 1999, pp. 205–208.

Clay, Ellen "The Class Mission Statement" In *Assessment Practices in Undergraduate Mathematics*, Bonnie Gold, et al., eds. Washington DC, Mathematical Association of America, 1999, pp. 155–157.

Dwyer, Carol A. et. al., *What is Quantitative Reasoning?: Defining the Construct for Assessment Purposes*. Educational Testing Service, Princeton, NJ: to appear.

Emert, John W. "Improving Classes with Quick Assessment Techniques." In *Assessment Practices in Undergraduate Mathematics*, Bonnie Gold, et al., eds. Washington DC, Mathematical Association of America, 1999, pp. 94–97.

Gold, Bonnie, Sandra Z. Keith, and William A. Marion, eds. *Assessment Practices in Undergraduate Mathematics*. MAA Notes, Vol. 49. Washington, DC: Mathematical Association of America, 1999. www.maa.org/saum/maanotes49/index.html.

Keith, Sandra Z. "Creating a Professional Environment in the Classroom." In *Assessment Practices in Undergraduate Mathematics*, Bonnie Gold, et al., eds. Washington DC, Mathematical Association of America, 1999, pp. 98–100.

Krawczyk, Donna and Elias Toubassi. "A Mathematics Placement and Advising Program." In *Assessment Practices in Undergraduate Mathematics*, Bonnie Gold, et al., eds. Washington DC, Mathematical Association of America, 1999, pp. 181–183.

Martin, William O. and Steven F. Bauman. "Have Our Students with Other Majors Learned the Skills They Need?" In *Assessment Practices in Undergraduate Mathematics*, Bonnie Gold, et al., eds. Washington DC, Mathematical Association of America, 1999, pp. 209–212.

Michael, Mark "Using Pre- and Post-Testing in a Liberal Arts Mathematics Course to Improve Teaching and Learning," In *Assessment Practices in Undergraduate Mathematics*, Bonnie Gold, et al., eds. Washington DC, Mathematical Association of America, 1999, pp. 195–197.

Olin, Robert and Lin Scruggs. "A Comprehensive, Proactive Assessment Program." In *Assessment Practices in Undergraduate Mathematics*, Bonnie Gold, et al., eds. Washington DC: Mathematical Association of America, 1999, pp. 224–228.

Poiani, Eileen L. "Does Developmental Mathematics Work?" In *Assessment Practices in Undergraduate Mathematics*, Bonnie Gold, et al., eds. Washington DC, Mathematical Association of America, 1999, pp. 202–204.

Sons, Linda R. "A Quantitative Literacy Program." In *Assessment Practices in Undergraduate Mathematics*, Bonnie Gold, et al., eds. Washington DC, Mathematical Association of America, 1999, pp. 187–190.

Sons, Linda R. et al. *Quantitative Reasoning for College Graduates: A Supplement to the Standards.* Washington, DC: Mathematical Association of America, 1996. www.maa.org/past/ql/ql_toc.html.

Steen, Lynn A. ed. *Mathematics and Democracy: The Case for Quantitative Literacy.* Princeton, NJ: National Council for Education and the Disciplines, 2001.

Steen, Lynn A. ed. *Why Numbers Count: Quantitative Literacy for Tomorrow's America.* New York, NY: The College Board, 1997.

Assessing Introductory Calculus and Precalculus Courses

Ronald Harrell and Tamara Lakins
Department of Mathematics
Allegheny College
Meadville, PA
rharrell@allegheny.edu
tlakins@allegheny.edu

Abstract. With the goal of improving student learning, Allegheny College assesses the effectiveness of its introductory calculus and precalculus courses by analyses of grade data, conversations with client departments, and information regarding such courses at similar institutions. The initial assessment led to substantial revisions in its offerings.

Background and goals: What did we hope to accomplish?

Allegheny College, a national liberal arts college located in Meadville, PA, has an enrollment of approximately 1800 students. The mathematics department teaches approximately 550 students per year in its introductory calculus and precalculus courses, which from 1990 to 2003 consisted of Math A (Intermediate Algebra), a one-year sequence Math 155/156 (Calculus/Precalculus), and a traditional calculus course Math 160 (Calculus I).

The original intent of Math A was to prepare students for the sequence Math 155/156. It also became a specific prerequisite for the following courses in client departments: introductory chemistry, introductory computer science, and the research design and statistics course in the psychology department. The original intent of the sequence Math 155/156 was to provide an alternate entry point into the regular calculus sequence for students with weaker precalculus backgrounds. In particular, the sequence covered selected precalculus topics in addition to the calculus topics traditionally covered in Calculus I and was designed to prepare students for Calculus II.

At the beginning of the fall semester, entering first-year students are placed in a mathematics course based on an algebra-trigonometry-precalculus-based placement exam and/or consultation with a member of the mathematics department. The placement is a non-binding recommendation.

The goals of our assessment project were to determine whether the intermediate algebra course and precalculus/calculus sequence were addressing the needs of our students and to make any needed changes to the courses, which in any case would include the addition of a regular assessment program. The intermediate algebra course had existed for over 20 years, and the precalculus/calculus sequence had been used for 13 years. In neither case had we ever assessed their effectiveness, and we had anecdotal evidence from members of the mathematics department that the courses were not preparing students at the level we expected for subsequent courses. It was time to take a hard look at both.

Description: What did we do?

During the 2001–02 academic year the mathematics department assessed the effectiveness of Math A (Intermediate Algebra) and Math 155/156 (Precalculus/Calculus) by reviewing data regarding student performance in these and subsequent courses, by conducting conversations with faculty in client departments, and by reviewing information about precalculus and introductory calculus offerings at the 26 colleges and universities identified by Allegheny College as in our comparison group.

Data on the distribution of grades were examined for the 359 entering first-year students who enrolled in Math A during a fall semester from 1997 to 2000 and who went on to take introductory chemistry, Math 155, introductory computer science, or the psychology statistics course. We omit the discussion of students who took the introductory computer science course, since their number was too small to draw viable conclusions. In addition, a second comparison of grade distributions was made for students taking these subsequent courses, based on mathematics placement level and regardless of whether the students took Math A. For this second comparison, we considered grades in Fall 1997 through Spring 2001, and the sample consisted of students who entered Allegheny in a fall semester from 1997 to 2000.

We examined data on the distribution of grades for the 310 entering first-year students who enrolled in Math 155 in a fall semester from 1997 to 1999 and who subsequently took Math 156 (the second course in the precalculus/calculus sequence) and Math 170 (Calculus II) at Allegheny College. These latter students were compared to the 224 entering first-year students who began in Math 160 (Calculus I) in a fall semester from 1997 to 1999 and who subsequently took Math 170 at Allegheny.

We obtained all grade data from the Allegheny College Registrar in electronic format. Student names were removed from the data, and fictitious identification numbers were used in order to ensure student anonymity.

We consulted 1–2 faculty members (typically the chair or faculty who teach courses requiring quantitative or mathematical skills) in our client departments regarding Math A and our calculus offerings. Client departments include biology, chemistry, computer science, economics, environmental science, geology, mathematics, physics, and psychology. In the case of Math A, we wished to specifically learn why Math A was a prerequisite for the four earlier mentioned courses, and which skills currently taught in Math A were considered essential for these courses. In the case of calculus and those departments that require it for their major, we wanted to know specifically why calculus was required and which skills were considered essential.

Data from the colleges and universities in our comparison group were obtained by researching the college and university web sites; when necessary, details were clarified by email and phone calls to the appropriate department chair, or another faculty member designated by a department chair. In particular, we wished to determine the following information

about each school in our comparison group: its precalculus and introductory calculus course offerings and the mechanisms by which it enables students with deficiencies in their mathematics preparation to prepare for calculus.

Initially, we planned to review the high school backgrounds of a sample of our students. This would have required reading folders of individual students to determine which math courses had been taken and which grades had been obtained. Organizing that information would have been complicated by the variety of math courses now taught in high schools. So given our resources, we decided to drop this part of the study.

Insights: What did we learn?

The analysis of the data, on student performance in Math A (Intermediate Algebra) and the courses for which it became a prerequisite, indicated that Math A did not seem to adequately prepare students for any of the subsequent courses, except possibly the psychology statistics course. (Year-by-year details of all the grade information discussed here can be found on MAA Online.[1]) In particular, of the 359 entering first-year students who enrolled in Math A during their first semester at Allegheny College, 246 (69%) earned a successful grade of C or higher. Of these 246 students, 112 went on to take introductory chemistry with 74 (66%) earning a grade of C or higher, 183 took Math 155 with 111 (61%) earning a grade of C or higher, and 54 took the psychology statistics course with 46 (85%) earning a grade of C or higher (see table below).

How Did Successful Math A Students Do In Successor Courses?

Successor Course	C or higher in successor course
Introductory Chemistry	66%
Math 155	61%
Psychology statistics	85%

Even more striking was that our second comparison showed that, of all students who enrolled in introductory chemistry or Math 155 from Fall 1997 to Spring 2001, students who placed in Math A and who did not take it prior to enrolling in one of these courses (presumably because they felt, despite placement test results, that they were already adequately prepared) often fared as well as, or better than, students who placed in Math A and took it first. In particular, 49 of the 74 students (66%) who placed in Math A but did not take it or Math 155 prior to enrolling in introductory

---

[1] www.maa.org/saum/cases/Allegheny_A.html

chemistry earned a grade of C or higher in chemistry, while 85 of the 141 students (60%) who placed in Math A and took it or Math 155 prior to taking introductory chemistry earned a grade of C or higher in chemistry. In the case of Math 155, 67 of the 107 students (63%) who placed in Math A but did not take it prior to taking Math 155 earned a grade of C or higher in the course, while 124 of the 231 students (54%) who placed in Math A and took it prior to taking Math 155 earned a grade of C or higher in the course. (See table below.) Thus, it appeared that Math A was ineffective in preparing students for subsequent courses.

Percent receiving C or higher among those placed into Math A

	In Introductory Chemistry	In Math 155
Took Math A first	60%	54%
Skipped Math A	66%	63%

The analysis of the data on student performance in the one-year precalculus/calculus sequence (Math 155/156) indicated that a smaller number of these students went on to take Calculus II (Math 170) than one might expect. Of the 310 entering first year students who enrolled in Math 155 during a fall semester from 1997 to 1999 only 52 (17%) went on to take Calculus II (either at Allegheny or elsewhere) by Spring 2001, while of the 446 entering first year students who enrolled in Math 160 during the same semesters, 248 (56%) went on to take Calculus II (either at Allegheny or elsewhere) by Spring 2001. Furthermore, students in Allegheny's Math 170 who began in Math 155/156 were generally not as successful as those who began in Math 160. Of the 47 first-year students who began in Math 155 during a fall semester from 1997 to 1999 and who went on to take Math 170 at Allegheny by Spring 2001, 29 (62%) earned a grade of C or higher. On the other hand, of the 224 students who began in Math 160 and went on to take Math 170 at Allegheny by Spring 2001, 168 (75%) earned a grade of C or higher. While the sample size is too small to be sure the results aren't merely chance variations, it appeared that the Math 155/156 sequence was not serving students as well as we had hoped.

The conversations with faculty in client departments revealed that Math A was serving more than one purpose. The chemistry and mathematics departments required Math A, or placement out of Math A, as a prerequisite for entry level courses which require students to possess traditional algebra skills. However, faculty in the biology, computer science, environmental science, and psychology departments emphasized wanting their students to possess general quantitative skills (such as good problem solving skills, the ability to translate word problems into an appropriate math-

ematical model, being able to work with and interpret data, and having good number sense), rather than specific kinds of algebra skills. Clearly at the level below calculus, something more than a course that only reviews algebra was needed.

In the case of Math 155/156 the chemistry, mathematics, and physics departments expected a thorough treatment of the concepts of calculus and a good knowledge of computational skills. The other client departments required at least some calculus for their major programs and expected students to be able to understand and apply the concepts of calculus. Often these same departments wanted calculus courses to be less theoretical and were primarily interested in having their students learn how to do only the more elementary computations. Thus there were two kinds of clientele for calculus courses.

Finally, an examination of course offerings at the 26 schools in our comparison group indicated that we were only one of two institutions that offer a course at the level of Math A, and only two institutions offer a course at the level of college algebra. On the other hand, a total of 15 colleges offer precalculus, seven offer a combined precalculus and calculus sequence similar to our Math 155/156, and five colleges offer no course which directly prepares students for calculus. Finally, eight institutions offer a one-semester alternative to the traditional first calculus course (usually a course which emphasizes applications from the social and/or life sciences).

Redesigning: What did we do?

The above findings indicated that our lower level course offerings were not diverse enough to meet students' needs, and some courses did not accomplish their intended purpose. After much discussion, the mathematics department replaced Math A (Intermediate Algebra) and Math 155/156 (Precalculus/Calculus), which were hierarchically designed to prepare students for the regular calculus sequence, with four courses that provide students with three options for beginning the study of mathematics, depending on their individual goals. The new courses are briefly described below; more detailed descriptions and course goals can be found on *MAA Online*.[2] Important in the design of the new courses was meeting the needs expressed by client departments.

The first option, Math 110 (Elementary Mathematical Models), replaced Math A. Math 110 is an elementary algebra-based modeling course that emphasizes the study of real world problems and models, and rates of change. Algebra is

---

[2] www.maa.org/saum/cases/Allegheny_B.html

reviewed as needed. The course is for those students who need a mathematics course but not a calculus course. The intended audience consists of humanities and social science students, who take it to fulfill a graduation requirement or who find it useful in a major field, such as economics, environmental studies, political science, or psychology.

The second option, Math 150 (Precalculus), is a standard college level course on the subject, intended only for those students who need to take the regular calculus sequence, but who also need to brush up on precalculus topics before doing so. Topics covered in the course were formerly taught in the sequence Math 155/156.

The third option is the sequence Math 157/158 (Calculus I and II for Social/Life Sciences), which replaces the Math 155/156 sequence. The sequence is for those students who need calculus, but not the thorough and more rigorous treatment presented in the regular calculus sequence. The emphasis is on the concepts of calculus and how they occur in problems from the life and social sciences. Topics in both single and multivariate calculus are covered. This option serves primarily biology, economics, and environmental science students.

The mathematics placement exam, which was previously used to determine placement in either Math A, Math 155, or Math 160, is still used by the department. The department now requires a particular score on that exam in order to recommend placement into Math 160. Students who do not achieve the target score may enroll in Math 110, Math 150, or Math 157, depending on high school background, confidence, and intended major. These latter three courses have no formal prerequisite. While Math 110 is a terminal course and is not intended to prepare students for Math 157, some students may opt to take it before attempting Math 157.

Students who have already received college credit for a calculus course may not take Math 110, 150, or 157 for credit. Furthermore, students who begin in the Math 157/158 sequence and later change to a major requiring the ordinary calculus sequence Math 160/170 are treated on an individual basis. A student may take Math 160 for credit after receiving credit for Math 157 but not after receiving credit for Math 158. (Such a student may still take Math 160, but will not receive credit toward graduation for the course.) Students who wish to take Math 170 after Math 158 receive individual advising.

In addition to creating these new courses and options, we also created a way to have an ongoing assessment of each in order to monitor their effectiveness. Content goals for each course are assessed using selected questions on final exams to gauge how well students have mastered the material. For the goal

- students will be able to communicate mathematical information in written form,

which pertains to each of the new courses, as well as the goal

- students will be able to choose, implement, refine, and interpret appropriate mathematical models for various real-world problems,

which pertains to Math 110 and the sequence Math 157/158, the assessment consists of short writing assignments, projects, or appropriate homework where writing is emphasized. Thus the assessment data consists of scores on selected final exam questions and instructors' impressions of the writing assignments and/or projects during the semester.

At the end of each semester the instructors for each course meet briefly to review and discuss the assessment data for that semester. They then submit a report of their findings to the department chair, who makes the contents of the report available to the entire department. It is hoped that, by reviewing a small amount of assessment data each year, the department will be able to maintain an ongoing and accurate picture of the effectiveness of the courses and our assessment methods, while at the same time not placing too great a burden on the faculty. Periodically, perhaps every three to five years, we will do a wider assessment, similar to the one reported here, that determines how well these courses prepare students for subsequent courses, not only in mathematics, but in other areas.

Other Comments

After we compiled and analyzed the data, the department spent several weeks of intense discussion creating the above four replacement courses. The two major sticking points were the exact nature of the replacement courses and finding a reasonable ongoing assessment plan for each course. Some faculty questioned whether ongoing assessment plans were needed, and getting them to see the usefulness and benefits of such plans was a hard sell. To help students and advisors, the department also made up documents explaining the new courses in detail and indicating which courses would benefit which students.

Acknowledgements. The authors wish to acknowledge the support and assistance of Allegheny College's Associate Dean Richard Holmgren in planning and implementing this assessment project. Funds from Dean Holmgren's Arthur Vining Davis Foundation grant supported the 2001–02 salary of Tamara Lakins, as well as the travel costs of both authors to three SAUM Assessment in Undergraduate Mathematics workshops. We also acknowledge the assistance and advice of our workshop team leader Bonnie Gold.

Mathematics Assessment in the First Two Years

Erica Johnson, Jeffrey Berg,
David Heddens
Department of Mathematics
Arapahoe Community College
Littleton, CO
erica.johnson@arapahoe.edu
jeff.berg@arapahoe.edu
david.heddens@arapahoe.edu

Abstract. The Arapahoe Community College Mathematics Department participated in a two-year college-wide assessment effort. During both years, the Department analyzed data from a common final exam in College Algebra. In the second year, the Department linked an entrance exam to the common final exam to measure student learning during College Algebra. These efforts laid the foundation for a continuing annual assessment process with an ultimate goal of assessing all program-level student-learning outcomes at least two ways.

Background and Goals

For several years the department has used a Calculus Readiness placement test produced by the Mathematical Association of America (MAA) as a common final for College Algebra. The department curved test scores based on the performance of all students taking the test and the curved grade counted in each student's overall course grade. Beyond curving the test scores, detailed analyses of student performance and/or feedback channels for improving pedagogy at the department level did not exist. The department felt that the test could be used as a starting point for its efforts to develop a discipline-level assessment program with the goal of developing an instrument useful for improving the department's ability to address its student learning outcomes.

The department has used the MAA Basic Algebra placement test as an entrance exam in College Algebra for a retention improvement project. As another component of its assessment efforts, the department decided to link the entrance exam and the common final using Colorado State Core Transfer Program student learning outcomes to measure student learning in ability to work with mathematical formulas and word problems.

Description

The department developed a mission statement in harmony with the College Mission statement and the current direction of the College:

> The mission of the Mathematics Department at Arapahoe Community College is to provide learning-centered mathematics education to students. The department offers courses for both full-time and part-time students supporting both transfer and career opportunities. The department is committed to quality learning-centered mathematics education valuing traditions and incorporating current effective pedagogical trends in the discipline, appropriate technology, and assessment of student learning.

Next, the department developed five student learning outcomes drawing heavily from *Crossroads in Mathematics: Standards for Introductory College Mathematics Before Calculus* [1]:

- Students will acquire the ability to read, write, listen to, and speak mathematics.
- Students will demonstrate a mastery of competencies identified by the competency-based syllabi for specific courses.
- Students will use appropriate technology to enhance their mathematical thinking and understanding and to solve mathematical problems and judge the reasonableness of their results.

- Students will engage in substantial mathematical problem solving.
- Students will acquire the ability to use multiple approaches—numerical, graphical, symbolic, and verbal—to solve mathematical problems.

Nearly all departmental courses address each of these outcomes. The wording of the second learning outcome was chosen to make use of the Colorado Community College System Core Transfer Program and Common Course Numbering System curriculum requirements. Both the Core Transfer Program and the Common Course Number System define student learning outcomes for individual courses offered by the department.

To test student performance on these learning outcomes, the Mathematics Department chose to use a Calculus Readiness placement test produced by the MAA as a common final for College Algebra. The department used the standard Calculus Readiness placement test in the 2001–02 academic year, but switched to the Calculator-Based Calculus Readiness placement test in the 2002–03 academic year because of a requirement that students have and use graphing calculators in College Algebra. The common final was given on the last day of classes and counted at least 10% of each student's overall grade. Student performance data was collected for both the fall and spring semester.

After mapping questions to Core Transfer Program learning outcomes, the department decided to perform the following detailed analyses:

- Student performance on common final, which pointed to strengths and weaknesses in departmental pedagogy.
- Longitudinal comparisons of overall student common final performance for academic years 2001–02 and 2002–03, which indicated degree of consistency of results between semesters in an academic year and between academic years.
- Comparison of student performance on linked questions between entrance exam and common final exam as a measure of student learning in ability to work with mathematical formulas and word problems over the semester.

Student Performance on Common Final

During the fall semester 2002, 158 students took the common final exam; the overall performance is presented in Figure 1. During the spring semester 2003, 180 students took the common final exam; the overall performance is presented in Figure 2. Table 1 presents in descending rank order students' overall performance by question for spring 2003. The third column in Table 1 specifies for each question the associated Colorado Core Transfer Program learn-

ing outcomes. Those associated with questions with correct response rates above 50% indicating areas of departmental pedagogical strengths are:

B. Perform algebraic manipulations including working with exponents, radicals, polynomial operations, factoring and algebraic fractions.

E. Work with formulas including formula evaluation and solving a formula for any of the variables.

F. Read and analyze problems in the form of word problem applications and obtain solutions using equations.

G. Solve first degree inequalities, higher degree inequalities and inequalities involving absolute value.

H. Recognize and graph linear functions, rational functions, absolute value functions, and graph inequalities in two variables.

I. Work with function notation and demonstrate knowledge of the meaning "function."

J. Demonstrate an understanding of function composition, one-to-one functions and inverse functions.

K. Examine, evaluate and graph exponential functions.

Core Transfer Program learning outcomes associated with questions with correct response rates below 50% indicating areas of departmental pedagogical weaknesses are:

F. Read and analyze problems in the form of word problem applications and obtain solutions using equations.

I. Work with function notation and demonstrate knowledge of the meaning "function."

M. Work problems and solve equations containing exponential and logarithmic functions.

O. Use at least two of the following techniques to solve linear and non-linear systems of the equations: substitution, addition, Gaussian elimination, Cramer's rule.

U. Work with series notation and sequence formulas, and counting principles.

Longitudinal Analyses

Comparison of fall and spring student performance data for the two academic years indicated no statistical difference between performance data for different semesters within an academic year.

The department changed from the Calculus Readiness placement exam to the Calculator Based Calculus Readiness placement exam prior to its 2002–3 assessment cycle. Since the exam version changed between spring 2002 and spring 2003, Colorado Core Transfer Program student learning outcomes linked exam questions. Other than question renumbering, 18 out of 20 questions on the two versions of the common final either were the same (3 questions) or tested the same outcome using slight wording and/or number variations (15 questions). The remaining 2 out of 20 questions on each version tested different outcomes and were not linked. Overall, the spring 2002 and spring 2003 distributions differed substantially, with a $p < 0.0001$. A "by question" comparison of spring 2002 and spring 2003 final exam correct response rates measured differences at the $\alpha = 0.03$ level of significance. From spring 2002 to spring 2003 correct response rates went up on 2 questions, went down on 6 questions, and remained statistically the same on 10 questions.

Rather than pursuing further study of these results, the department decided to remain with the same version of the common final from the 2002–3 to the 2003–4 assessment cycle to produce more reliable data to measure change between academic years.

Comparison between Entrance Exam and Common Final Exam

The department targeted two Core Transfer Program student learning outcomes,

E. Work with formulas including formula evaluation and solving a formula for any of the variables, and

F. Read and analyze problems in the form of word problem applications and obtain solutions using equations,

because of their relevance across department curricula and their harmony with the overarching departmental student learning outcomes. These outcomes tied the entrance exam and the common final to provide a means to measure student learning. Two questions on the entrance exam and three questions on the common final addressed these outcomes. Each student had a correct response count of 0, 1, or 2 on

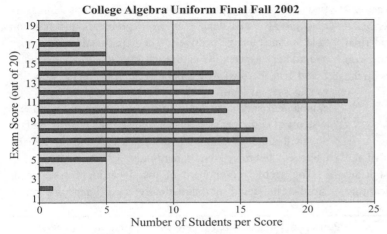

College Algebra Uniform Final Fall 2002

Figure 1. Distribution of Student Performance on Common Final Exam, Fall 2002.

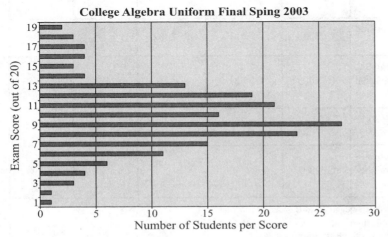

College Algebra Uniform Final Sping 2003

Figure 2. Distribution of Student Performance on Common Final Exam, Spring 2003.

Question	% Correct	Learning Outcome
19	88.33%	H
11	79.44%	K
15	79.44%	I
3	74.44%	I
17	72.22%	G
9	70.00%	E
18	69.44%	F
16	67.22%	H
4	63.89%	E
7	63.33%	H
12	63.33%	B
14	62.78%	B
5	57.78%	J
6	51.67%	H
10	48.33%	U
20	43.89%	F
13	42.78%	I
1	41.11%	O
8	41.11%	M
2	36.11%	F

Table 1. Rank Order of Student Performance on Questions on College Algebra Common Final Exam — Spring 2003

the entrance exam and 0, 1, 2, or 3 on the common final exam. The mean Fall 2002 and Spring 2003 entrance exam competency measurements were 1.52 (standard deviation 0.62) and 1.40 (standard deviation 0.66) respectively. The mean Fall 2002 and Spring 2003 final exam competency measurements were 1.71 (standard deviation 1.05) and 1.44 (standard deviation 1.00) respectively.

Investigation of degree of change in student competency between entrance and final exam was performed by computing the difference $d = p_2 - p_1$, where p_1 = proportion of correct responses on entrance exam and p_2 = proportion of correct responses on final exam. The mean value of d for Fall 2002 was -0.19 (standard deviation 0.41) and for Spring 2003 was -0.22 (standard deviation 0.41). Negative values of d reflect a reduction in relative rate of correct responses between entrance and final exam.

Contingency tables classified students by their entrance and final competency measures for Fall 2002 and Spring 2003. Concordance measure gamma for Fall 2002 was 0.29 and for Spring 2003 was 0.31. Hence, each semester has a limited degree of concordance between entrance and final competency, implying a slight tendency for higher scores on the entrance exam to be associated with higher scores on the final exam.

Insights

Improvement of student learning is the foundation of the assessment process. Rather than being a single event at the "end of the process", changes that improve student learning happen continuously throughout an effective assessment cycle. Refinement of the process, increased student awareness of expectations and involvement in assessment activities, constant faculty discussions of ineffective and/or improved pedagogical methods and subsequent decisions to modify pedagogy, and administrative support of program assessment efforts are some components of the assessment cycle that are integral to achieving the goal of improved student learning.

In the first of three components of the departmental assessment process, student common final correct response ranges indicate departmental strengths in Colorado Core Transfer Program student learning outcomes, B (63%, 63%), E (70%, 64%), F (69%), G (72%), H (88%, 67%, 63%, 52%), I (79%, 74%), J (58%), and K (79%) and departmental weaknesses in Colorado Core Transfer Program student learning outcomes, F (44%, 36%), I (43%), M(41%), O (41%), and U (49%). Learning outcomes F and I are identified as both strengths and weaknesses. Questions addressing outcomes F and I dealt with word problems and function notation/meaning respectively. The department will identify and address aspects indicating weaknesses in competencies F and I in the next assessment cycle. Questions indicating weaknesses in learning outcomes M, O, and U dealt with exponential and logarithmic equations, linear and nonlinear systems of equations, sequences, and series. Department faculty will discuss strengths and weaknesses and modify the College Algebra syllabus accordingly.

The second component of the department assessment process indicated no statistically significant difference

Outcome/Tool	Project or Portfolio	Standardized Exam	Pre-test/ Post-test	Faculty Survey	Student Survey
Students will acquire the ability to read, write, listen to, and speak mathematics.	CP2003				
Students will demonstrate a mastery of competencies identified by the competency-based syllabi for specific courses.	CP2003	CF2001	PP2002		
Students will use appropriate technology to enhance their mathematical thinking and understanding and to solve mathematical problems and judge the reasonableness of their results.	CP2003	CF2001			
Students will engage in substantial mathematical problem solving.	CP2003				
Students will acquire the ability to use multiple approaches—numerical, graphical, symbolic, and verbal—to solve mathematical problems.	CP2003				

CF2001 — College Algebra common final data collection and analysis began in 2001–2 academic year
PP2002 — College Algebra entrance exam/common final data collection and analysis began in 2002–3 academic year
CP2003 — Calculus I common projects with scoring rubric beginning in 2003–4 academic year

Table 2. Assessment Methods Used to Measure Student Learning Outcomes

between the data from fall semester to spring semester during an academic year. The department has decided to collect another year of data to confirm the trend. If confirmed, the department will have the choice of either combining semester data into yearly data to increase statistical power or analyzing single semester data in an academic year to reduce resource requirements.

The third component of the department assessment process indicated a decrease in student performance on the two targeted Colorado Core Transfer Program student learning outcomes and a slight dependency of final competency on entrance competency. The decrease could be partly due to the fact that the two entrance exam questions were different from the three common final exam questions. The department decided to improve the method in 2003–4 by revising the entrance exam to include the three questions from the common final. Data collected using identical questions will allow statistically sound t-tests of proportions, contingency analysis, and logistic regression to better measure student learning during the course of the semester.

In its 2003–4 assessment cycle, the department will introduce an additional component to its assessment process. This additional component will be two common projects with associated scoring rubric in Calculus I. Students will receive a copy of the rubric with the project assignment. This component will assess all five discipline-level student learning outcomes and move the department toward the ACC Assessment Committee goal of assessing all program-level student learning outcomes at least two ways by the 2004–5 academic year. Table 2 documents the department progress towards this Assessment Committee goal.

Acknowledgments

Department members have been aided by participation in the Mathematical Association of America's project Supporting Assessment in Undergraduate Mathematics (SAUM). Members of the department attended workshops in Burlington, Vermont in July 2002 and Baltimore, Maryland in January 2003, presented at a special SAUM session in Boulder, Colorado in August 2003, and participated in the SAUM poster session and the culmination of the workshop #2 series in Phoenix, Arizona in January 2004. Professional contacts and interchange with faculty across the nation involved in similar assessment activities and expert advice have been helpful in developing the ACC Mathematics Department assessment process. Further feedback from the SAUM publication effort will be incorporated into the department's assessment process.

Bibliography

1. *Crossroads in Mathematics: Standards for Introductory College Mathematics Before Calculus*. Memphis, TN, American Mathematical Association of Two-Year Colleges, 1995.

Using Assessment to Troubleshoot and Improve Developmental Mathematics

Roy Cavanaugh, Brian Karasek,
Daniel Russow
Department of Mathematics
Arizona Western College
Yuma, AZ
roy.cavanaugh@azwestern.edu,
brian.karasek@azwestern.edu,
daniel.russow@azwestern.edu

Abstract. In the fall of 2003 and spring of 2004, the Arizona Western College Mathematics Department assessed preparedness of students entering our beginning algebra and intermediate algebra courses. We examined how student preparedness was correlated with how they entered the course (prerequisite course, placement test, or by instructor sign-in) and with the length of time since they took their last mathematics course.

Background

Arizona Western College is a rural community college located in southwestern Arizona. Its service area spans from the Mexican border 150 miles north to La Paz County. There are approximately 5000 full-time students enrolled at the college, with a total of around 10,000 students.

The mathematics department at Arizona Western College has 13 full-time instructors and approximately 30 part-time instructors. Our developmental mathematics program consists of three courses: Mathematics Essentials (MAT 072), Beginning Algebra (MAT 082) and Intermediate Algebra (MAT 122). The developmental mathematics program is meant to prepare students to take our college-level mathematics courses. Students taking developmental mathematics courses are initially placed into the appropriate course via the college placement test; entrance into the later two courses can also be by passing the previous course or by instructors signing students in. Approximately 1500 students are enrolled in developmental mathematics courses each semester at Arizona Western College.

The mathematics department was concerned that students entering our beginning algebra course appeared not to be as prepared as instructors would have hoped. Some instructors who teach these courses stated that even those who took our pre-algebra course did not seem to possess the skills necessary to be successful in beginning algebra. We decided to assess student preparedness in both our beginning and intermediate algebra courses to see if the students possessed the skills that we, as a department, felt were vital to their success in the course.

Since students enter these courses in one of three ways (via prerequisite course, placement test, or by instructor sign-in), we also felt it was vital for us to assess how they performed in relation to how they gained entrance into the course. So, in essence, we are assessing our placement process as well as how prepared students are when they are signed into a course. We also decided to assess how well students are prepared in relation to the time elapsed since their last mathematics course.

Description

To begin, we polled instructors on what skills they felt were vital to a student's success in each of these two algebra courses. We have split the learning objectives for algebra into these two courses. Intermediate algebra picks up where beginning algebra left off. One book is used for both courses. After getting instructor input, reviewing the objectives for these courses on our official course sylabi, and seeing what skills were assumed by the textbook that we are using, we arrived at a set of competencies that we felt were vital for student success in each of these courses.

The prerequisite skills for beginning algebra included operations with integers, operations with fractions, percents, ratios and proportions, order of operation, evaluating a formula, solving a simple linear equation, and the Pythagorean Theorem. In intermediate algebra, the prerequisite skills included solving linear equations, factoring, operations with polynomials, solving a system of two equations and two variables, finding the slope of the line passing through two points, and graphing a line.

We then developed our assessment tools for each of these courses. Each assessment tool consisted of ten multiple choice questions, numbered 3 through 12, along with two background information questions number 1 and 2 (See Appendix A). The background information included how students gained entrance into the courses (via prerequisite course, placement test, or by instructor sign-in) and the length of time since their last mathematics course. The ten mathematics questions were developed so that only one competency was being assessed by each question. This way, if students got the question wrong, we would be relatively certain that they got it wrong as a result of not having mastered a certain competency. We were successful in doing this for all but two of the questions.

Two questions caused some irregularities in our analysis. On our beginning algebra test, one question was intended to measure a student's ability to perform operations on an expression in the correct order. Unfortunately, due to an oversight on our part, the expression evaluated to a fraction approximately midway through the computation. Therefore we are concerned that we may not be able to determine which competency (order of operations or fraction skills) caused their incorrect response. Similarly, on the intermediate algebra test, one question was intended to measure their ability to graph a linear equation. Problems in printing caused the graphs to be difficult to read. Results from this question are therefore suspect.

Assessments were developed for grading via Scantron. We ran into a little difficulty sorting results in the fall assessment. Our results reflected this in the fact that we did not include in our analysis the time elapsed since their last math class. In fact, we had to sort by hand how they gained entrance to the course. In the spring, we developed our own Scantron form and we were able to then use a more advanced scanning machine which would allow us to analyze the data with the aid of our computers.

In both the fall and spring, assessments were given in all beginning and intermediate algebra classes on the second class meeting of the semester. Students were notified on the first class meeting that they would receive the test on the next class meeting. All instructors were encouraged to count the assessment as part of the student's semester grade. Using the assessment as part of the student's grade encourages the students to prepare and to give it appropriate effort. In total, 739 students were assessed in the fall and 763 students were assessed in the spring.

Insights

Our overall data for beginning algebra supported our initial hypothesis that students entering beginning algebra are not sufficiently prepared for the course (as measured by our test). We also found that there was no significant difference in preparedness based on method of entrance into course, time elapsed since their last mathematics course was taken, or semester administered.

The first graph in Appendix B summarizes the results for the assessment in our beginning algebra course for the fall of 2003 and spring of 2004 semesters. The competencies being measured appear below the graph. For example, fewer than fifty percent of all of the students enrolled in beginning algebra could add two simple fractions. Only four of the ten competencies measured had been mastered by more than sixty percent of our students.

Recommendations for change were made at the department level. Based upon the above findings the department recommended that a diagnostic placement test for our developmental mathematics courses be investigated with the intent of moving towards a modular competency based developmental program. Further recommendations are to implement a more seamless transition from pre-algebra to beginning algebra through the use of a single-author series of texts.

Our overall data for intermediate algebra indicated that students are better prepared for entrance into intermediate algebra than students entering beginning algebra. In intermediate algebra there were significant effects ($a = 0.05$) of the different methods of entrance and of the time elapsed since their last mathematics course was taken. Students performed best on the test if they gained entrance to the course via our college placement exam (ACCUPLACER™). Students entering via the prerequisite course or by instructor permission performed similarly. The time since their last mathematics class also was a determining factor in their performance on the test. Students without recent mathematics coursework ($n > 5$ years) performed at a lower level than students with more recent exposure.

The second graph in Appendix B summarizes the results for the assessment in our intermediate algebra course for the fall of 2003 and spring of 2004 semesters. Again, the competencies being measured appear below the graph. Only two competencies scored below sixty percent. As discussed earlier, we felt that question number 12 scored lower in the spring because of readability problems with the graphs on our new test form.

Recommendations for change were again made at the department level. Based upon the above findings the department recommended that students who have not taken a mathematics course within the past three years should be required to retake the placement exam. Implementation of this recommendation will require further discussion as well as institutional support.

One concern of our analysis is the quality of our test as an indicator of success in the course. We are planning to correlate student's scores on the test with their semester grade to determine if the test is an indicator of success in the course. This will entail collecting student grades and then re-analyzing the tests. To facilitate this process, in the fall of 2004 we will again perform the assessment, but this time we are going to ask the instructors to hold onto the forms until the end of the semester. They will then be asked to enter the final grade the students receive in the class. We can then determine if our objectives are an indicator of success in the course.

Acknowledgments

We would like to thank the MAA, SAUM, the Arizona Western College Math Department, and Dr. George Montopoli for his assistance in the statistical analysis of our data. Additional information regarding our assessment can be found on-line at www3.azwestern.edu/math/assessment/AWCAssessment.htm.

Appendix A. Assessment Tools for Beginning and Intermediate Algebra

Mat 82 - Beginning Algebra
Prerequisite Examination   Name

Background Information

1) You gained entrance to this course by

a) taking the prerequisite course	b) taking the placement test	c) instructor permission

2) How long has it been since you took your last mathematics course?

a) 0-1 year	b) 1-3 years	c) 3-5 years	d) over 5 years

Mathematics Questions

3) Perform the operations and simplify.	4) Add the fractions and simplify.
$-2-5-(-6)$	$\dfrac{2}{3}+\dfrac{1}{4}$
$a)\,13\quad b)\,-13\quad c)\,28\quad d)\,-1$	$a)\,\dfrac{11}{12}\quad b)\,\dfrac{3}{7}\quad c)\,\dfrac{3}{12}\quad d)\,\dfrac{2}{7}$
5) Divide the fractions and simplify.	6) If the formula for the volume of a box is $V = l\cdot w\cdot h$, find the volume when $l = 5,\ w = 2,\ and\ h = 3$.
$\dfrac{1}{5}\div\dfrac{2}{3}$	
$a)\,\dfrac{2}{15}\quad b)\,\dfrac{1}{4}\quad c)\,\dfrac{5}{3}\quad d)\,\dfrac{3}{10}$	$a)\,10\quad b)\,25\quad c)\,30\quad d)\,11$
7) One week, Jose made $350 for 18 hours of labor. How much money would Jose make for working a 40-hour week?	8) If a television sells for $325 and there is a 9.2% sales tax, how much tax would be paid?
$a)\,\$400.52\quad b)\,\$750.00$	$a)\,\$2.99\quad b)\,\$29.90$
$c)\,\$777.78\quad d)\,\$812.87$	$c)\,\$35.49\quad d)\,\$50.00$

9) Solve the equation for x.

$3x - 2 = 5$

a) $x = 21$ b) $x = 3$

c) $x = \dfrac{3}{5}$ d) $x = \dfrac{7}{3}$

10) Find the length of the missing side of the right triangle.

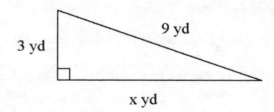

3 yd

9 yd

x yd

a) $\sqrt{72}$ yd   b) 6 yd

c) $\sqrt{90}$ yd   d) 18 yd

11) Perform the operations and simplify.

$9 \div 3 + 8 - 7 \times 2$

a) -3 b) 8   c) 10   d) 12

12) Perform the operations and simplify.

$5 - \left(2 - 4\right)^2 \div 3$

a) 3   b) $\dfrac{1}{3}$ c) $\dfrac{11}{3}$ d) 5

Mat 122 – Intermediate Algebra	
Prerequisite Examination	Name

Background Information

1) You gained entrance to this course by

a) taking the prerequisite course	b) taking the placement test	c) instructor permission

2) How long has it been since you took your last mathematics course?

a) Less than 1 year	b) 1 to less than 3 years	c) 3-5 years	d) over 5 years

Mathematics Questions

3) Solve the following equation for x.

$$3x - 2 = 4x + 5$$

$a)\ 1 \quad b)\ -7 \quad c)\ 7 \quad d)\ -1$

4) Solve the following equation for y.

$$6y + 3 = 21$$

$a)\ 4 \quad b)\ 7 \quad c)\ -4 \quad d)\ 3$

5) Factor the following expression completely.

$$x^2 - 7x + 10$$

$a)\ (x-5)(x+2) \quad b)\ (x+10)(x+1)$

$c)\ (x-5)(x-2) \quad d)\ (x+3)(x-7)$

6) Factor the following expression completely.

$$4xy^2 + 12x^2 y^3$$

$a)\ 4(xy^2 + 3x^2 y^3) \quad b)\ 4xy^2(1 + 3xy)$

$c)\ 4x^2 y^3(1xy + 3y) \quad d)\ 4xy(y + 12xy^2)$

7) Perform the indicated operations and simplify.

$$\left(4x^2 + 6x - 9\right) - \left(2x^2 + 12x - 5\right)$$

$a)\ 2(x-2)(x+1) \quad b)\ 2x^2 - 3x - 2$

$c)\ 2x^2 - 6x - 14 \quad d)\ 2x^2 - 6x - 4$

8) Perform the indicated operations and simplify.

$$\left(4xy^2\right)\left(3x^2 y\right)$$

$a)\ 12x^2 y^2 \quad b)\ 12x^3 y^3 \quad c)\ 12x^4 y^4 \quad d)\ 7x^3 y^3$

9) Perform the indicated operations and simplify.

$(4x+9)(2x-3)$

a) $8x^2 - 27$ b) $6x^2 + 6$

c) $8x^2 + 30x - 27$ d) $8x^2 + 6x - 27$

10) Solve the following system of equations.

$4x + 5y = -15$

$-2x + 5y = -45$

a) $\begin{aligned} x &= 30 \\ y &= 27 \end{aligned}$ b) $\begin{aligned} x &= 7 \\ y &= -5 \end{aligned}$

c) $\begin{aligned} x &= 5 \\ y &= -7 \end{aligned}$ d) $\begin{aligned} x &= 6 \\ y &= -37.5 \end{aligned}$

11) Find the slope, m, of the line passing through the following points:

$(-2,6)$ and $(3,-4)$ given that $m = \left(\dfrac{y_2 - y_1}{x_2 - x_1} \right)$

a) 2 b) $\dfrac{10}{7}$ c) -2 d) $\dfrac{7}{10}$

12) Select the proper graph for the following line.
$3x + 4y = 12$

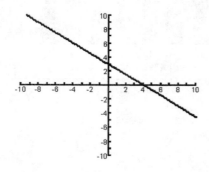

a)

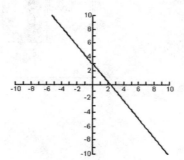

b)

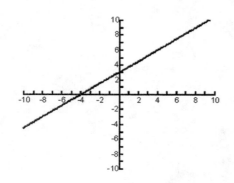

c)

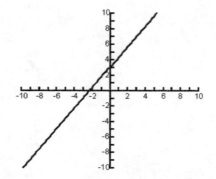

d)

Appendix B. Assessment Results for Beginning and Intermediate Algebra

Beginning Algebra Results:

Q.	Competency
3	Signed Number Operations
4	Addition of Fractions
5	Division of Fractions
6	Evaluation a Formula
7	Ratio Application
8	Percent Application
9	Solving a Linear Equation
10	Pythagorean Theorem
11	Order of Operations
12	Order of Operations

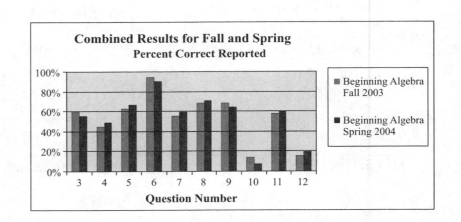

Intermediate Algebra Results

Q.	Competency
3	Solving a Linear Equation
4	Solving a Linear Equation
5	Factoring a Trinomial
6	Factoring Common Factors
7	Subtraction of Polynomials
8	Multiplication of Monomials
9	Multiplication of Binomials
10	Solving a System of Equations
11	Calculating Slope
12	Graphing a Linear Equation

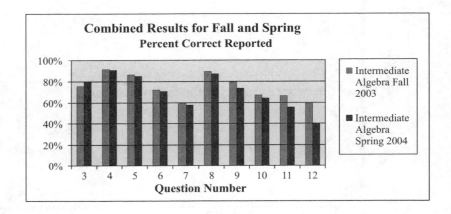

Questions about College Algebra

Tim Warkentin and Mark Whisler
Department of Mathematics
Cloud County Community College
Concordia, KS
tlwarkentin@hotmail.com
mwhisler@cloud.edu

Abstract. To begin assessing their College Algebra course at a small, rural community college in north-central Kansas, the authors investigated two questions. The first question was about the best format for the course, while the second concerned the effectiveness of the preceding course in the algebra sequence, Intermediate Algebra, in preparing students for College Algebra.

Background and Goals

Cloud County Community College is a small rural two-year college in north-central Kansas. The college serves a twelve county service area of over 8000 square miles on two widely separated campuses, and is an open enrollment college. Over the years we have seen a shift in our enrollments, as more and more of our students come to the college requiring one or more developmental courses. In fact, the vast majority of our enrollments, roughly 92% during the 2001–02 school year, are for classes at or below the level of College Algebra.

We naturally felt that our focus should be on these classes. We wanted to learn more about how our students fared when they moved from their developmental courses to College Algebra. In order to start small, we decided to limit this investigation to students who had taken the previous course in the sequence, Intermediate Algebra. We also wanted to investigate the best format for College Algebra itself.

Historically, Cloud County Community College has had the same problems as the rest of the nation in terms of student success in College Algebra. Thus, several years ago, a one-credit course, Explorations in College Algebra, was instituted as an attempt at intervention. This course was required of all students who took College Algebra at our Concordia campus. It was designed and implemented as a series of activities in which the students used graphing calculators and other technology, often in groups, to examine certain topics in more detail. These topics included basic regression analysis, the laws of gravity and light intensity, and a deeper investigation of exponential functions through applications to finance.

We wanted to determine whether the course fulfilled one of its primary goals: to increase performance in College Algebra. That data had never been gathered. We also received complaints on a regular basis from students because they were required to take a course with little or no perceived benefit to them. We were concerned that this requirement may have led to decreases in enrollment in main campus daytime College Algebra classes. Thus one issue we decided to examine was trying to find the best format for our College Algebra classes.

This effort formed the largest part of our project, but was not our original idea. As mentioned above, we were interested in whether our developmental classes provided a good preparation for College Algebra, and this was the idea we presented in our initial meeting in San Diego. Being new to the world of assessment, we had trouble deciding how we were going to measure this, and even what we were going to measure. We were asked to refine this idea. We decided that the examination of the best format for College Algebra would be our first project. This was particularly attractive since one member of our team (Warkentin) was teaching the three daytime sections of College Algebra in the spring semester. Thus we wouldn't have to worry about the effect of different instructors on any results.

Details of the assessment program

Since we were interested in finding the best format for the class, each section was delivered in a different format. Section A met three days a week, with an Explorations laboratory that met once a week for one and a half hours. Section B met each day of the week, with the Tuesday and Thursday sections meeting for 45 minutes each, without any lab activities. This time was instead used mainly to slow down the pace of lecturing so that students had more time to process and practice. The last section, Section C, met three days a week, without any additional time or intervention.

Because enrollments for the Spring semester had already taken place, there was little we could do to group students in any way. At the time of the project, students had to satisfy at least one of two prerequisites: two years of high school algebra with a C or better *and* appropriate test scores, or a grade of C or better in Intermediate Algebra, the previous class in our algebra sequence. Thus these classes had a fairly wide range of abilities and preparations. Admittedly this makes any conclusions from this study less reliable, but this was our starting point.

In each of these sections, we used questions on multiple-choice tests to measure student performance. Since these test questions were tied to objectives from the text, we used those as outcomes. We also tracked which section performed best on each chapter test, as well as the final examination.

While Warkentin investigated formats for College Algebra, the other member of the team (Whisler) looked into how well our developmental courses prepared students for College Algebra. With the help of our advisement office, we looked at the grades that students who took Intermediate Algebra received in College Algebra. This search went back through the past two years only, since we had a significant faculty turnover at that time.

Findings

Results of the investigation into the best format for College Algebra are partially summarized in tables that can be found on the SAUM web site.[1] We found that the section that met each day (Section B) performed the best on six of seven chapter exams. This result, though, did not carry over to the final exam, and in fact this section had the worst performance on the final exam. The next best performance during the semester came from the section that met three times per week (Section C), while the section with the Explorations lab (Section A) did worst on chapter exams. This section performed best on the final, however.

Table 1 displays success rates in College Algebra of students who took Intermediate Algebra from us and earned a grade of C or better. For the purposes of this study, we defined success to mean that the student earned a grade of C or better in the class the *first* time they took the class. That is, if a student had taken Intermediate Algebra, subsequently withdrawn from College Algebra, but then later succeeded in earning a grade of C or better, that was not considered a success. This is perhaps too stringent a criterion, since it doesn't take into account other life factors which may have had a significant impact on the student in a given semester.

	% of developmental students in class:	Success rate (%)
Fall 2000		
Section A	28.6 (5/21)	20.0.(1/5)
Section B	27.3 (6/22)	100 (6/6)
Section C	36.0 (9/25)	55.6 (5/9)
Overall	36.8 (20/68)	60.0 (12/20)
Spring 2001		
Section A	50.0 (11/22)	81.8 (9/11)
Section B	63.6 (14/22)	78.6 (11/14)
Section C	66.7 (8/12)	75.0 (6/8)
Overall	58.9 (33/56)	78.8 (26/33)
Fall 2001		
Section A	22.2 (6/27)	66.7 (4/6)
Section B	21.4 (3/14)	33.3 (1/3)
Section C	17.6 (3/17)	100.0 (3/3)
Overall	20.7 (12/58)	66.7 (8/12)

Table 1. Success in College Algebra of successful Intermediate Algebra students.

One message here seems to be that developmental students who take College Algebra in the fall are somewhat more at risk. It seems likely that many students did not keep up their studies or use what they had learned over the summer, and so they simply forgot a good portion of what they had

---

[1] www.maa.org/saum/cases/CCCC.html

learned. It is also the case that they are competing against incoming freshmen who are more talented mathematically. There are other factors that come into play; there are probably multiple reasons for this situation. On the positive side, the success rate was significantly higher in the spring. This may indicate that we do a good job in our developmental courses overall preparing our students for College Algebra.

We also looked at success rates of students in College Algebra organized by the grade they earned in Intermediate Algebra. This information is contained in Table 2.

	C	B	A
Fall 2000	0% (0/8)	50% (5/10)	100% (2/2)
Spring 2001	61.5% (8/13)	93.7% (15/16)	75% (3/4)
Fall 2001	0% (0/2)	83.3% (4/6)	100% (4/4)
Overall	34.8% (8/23)	75.0% (24/32)	90% (9/10)

Table 2. Success rates in College Algebra sorted by grade earned in Intermediate Algebra.

We can see from this table that, for instance, of the 20 developmental students who took College Algebra in the Fall 2000 semester, eight of them earned a C in Intermediate Algebra. Of those eight students, none of them succeeded, according to the definition of success given above. Overall, 23 students taking College Algebra over these three semesters earned a grade of C in Intermediate Algebra, and of those 23, eight of them succeeded. Similar interpretations should be made for the other entries.

It seems clear that students who received a C in Intermediate Algebra typically struggled in College Algebra. In fact, just under 35% of those students earned a grade of C or better in College Algebra, which is how we defined success in the class. None of them earned an A, and only two earned a B. It should be no surprise that students who were allowed into the class despite not fulfilling one of the prerequisites fared even worse. For this reason we decided not to track this group of students. Some of these students simply slipped through the cracks in the advising process, but some of them were highly motivated, typically non-traditional students. They provided the only successes in this category. Overall, roughly 63% of our students who took Intermediate Algebra in this two-year period earned a C or better in College Algebra.

Use of Findings and Next Steps

The change with the greatest impact is likely to be the change in format that we instituted in the fall of 2002 in College Algebra. While none of the results we found reached the level of statistical significance, we felt it was worth trying to combine the apparent advantages of daily classes on results of chapter tests, and the Explorations lab on final exam performance. We are offering all of our daytime sections of College Algebra as classes that, along with its companion class, College Algebra Explorations, meet every day. We plan to continue our evaluation of this format with a smaller set of outcomes, to see if it nets any gains in student performance.

We had already raised the level of performance necessary in Intermediate Algebra to move on to College Algebra to a grade of B or better, and we intend to hold to that standard, despite pressure. One consequence of this policy is that we have received requests for alternatives to Intermediate Algebra if a student has earned a C in that course. We also are looking for ways to implement early intervention for developmental students who take College Algebra in the fall in order to improve their likelihood of success.

Even though one of us (Warkentin) is a veteran teacher, we were both new to the process of using assessment techniques for our program. We both feel that this experience has been a valuable one, but we believe that it would be difficult to sustain at the level of this case study. We are continuing, so far, the study of how students who take Intermediate Algebra perform in College Algebra. The challenge, as always with a small school, is to keep up our attention level to assessment and make it part of our departmental culture.

Assessing the General Education Mathematics Courses at a Liberal Arts College for Women

Abdel Naser Al-Hasan and Patricia Jaberg
*Department of Mathematics and
Computer Science
Mount Mary College
Milwaukee, WI*
alhasana@mtmary.edu
jabergp@mtmary.edu

Abstract. At a private Catholic college in the Midwest, an assessment plan was developed to assess the effectiveness of introductory courses taken by students to fulfill a basic mathematical competency requirement. An attitude survey was developed to give us information regarding the impact of mathematics anxiety and other affective factors on students' mathematical skills. A subset of the mathematics competency examination was selected and administered after students completed a remedial mathematics course that fulfilled the college mathematics competency requirement. The results were compared to students' original responses to see if there was any improvement.

Background and Purpose

Mount Mary College, located in metropolitan Milwaukee, is Wisconsin's oldest Catholic college for women with approximately 600 full-time undergraduate, 600 part-time undergraduate, and 175 graduate students. The Department of Mathematics and Computer Science consists of four full-time and nine part-time faculty members. The department offers BS degrees in both Mathematics and Computer Science.

The overarching goal for the current assessment program at Mount Mary College is to assess the effectiveness of our remedial mathematics courses and advise the department and the larger college community on how to better serve students in fulfilling the mathematics competency requirement. The current competency requirement consists of either achieving a score of 70% on the competency test upon entrance or taking a remedial mathematics course. The competency test consists of 32 multiple choice items related to basic operations with whole numbers, integers and rational numbers, ratio and proportion, number sense, and basic probability and statistics. A student scoring below 47% on the competency test is placed in a developmental mathematics course. A student scoring between 48% and 69% is placed in a liberal arts mathematics course.

Initial examination of these courses included discussions with departments that require the algebra sequence, enrollment data in the developmental and the liberal art courses since 1998, materials used for these courses, and student course evaluations. The review of these courses revealed the following:

- Developmental Mathematics, a course that did not earn core credits but fulfilled the math competency requirement, did not adequately prepare students to continue in the algebra sequence. There was a gap between the mathematics in the developmental course and the mathematics expected to be learned in the Introductory Algebra course. Material covered in this course was mostly a review of basic operations with whole numbers, integers, rational numbers, percent, ratio and proportions. Although introduction to algebra was part of the curriculum, students' background limited the amount of time devoted to this topic.

- Mathematics for the Liberal Arts lacked clearly stated goals. This course was introduced three years ago as an alternative to Developmental Mathematics and had not been assessed since that time. The course content was established after polling other departments. In an attempt to meet the diverse mathematical needs as perceived by other departments, the course was a collection of disparate topics such as basic set theory, logic, and probability with limited exposure to real-life applications. Since this was recommended for students who only narrowly miss passing a mathematics competency examination, the need for an examination of the course goals in relation to the students' need for quantitative literacy became evident.

The initial examination of course goals and enrollment data led to the conclusion that in order to develop an effective assessment program, we needed to reexamine the learning goals for each of these two courses, with a possible revision and realignment of the curriculum to ensure proper articulation.

Developing the assessment program

Discussions during the initial phase for our Developmental Mathematics, Mathematics for the Liberal Arts, and Introductory Algebra courses focused on the following:

- Original course goals in relation to the overall departmental goals
- The existing curriculum in relation to what it is we want students to learn
- The currently existing competency and placement tests
- An assessment plan for the presently-existing courses

This initial phase required a great deal of discussion concerning the purpose of each course and how each course fits into a cohesive sequence. Examination of the enrollment records from the last five years revealed that only about 25% of students enrolled in these remedial courses continued into the algebra sequence. This led us to concentrate on those students who did not pass the competency test and placed into either Developmental Mathematics or Mathematics for the Liberal Arts. During this phase, we developed new goals and objectives for these two courses[1] and began examining materials that would reflect these new goals and objectives.

These curricular changes were scheduled to be in place by the fall 2003 semester. However, the assessment phase was begun immediately. A pilot assessment plan was used during fall 2002 and spring 2003 to help inform our decision-making for the future. The pilot assessment plan focused primarily on skills and attitudes. We are currently working on modifying the assessment plan to reflect the developing learning goals and curriculum changes. Since all students entering Mount Mary College take the competency test, we selected ten items from this test to create a post-course assessment test. These items were used for the Developmental Mathematics[2] and the Liberal Arts[3] courses. Some items were identical, but others reflected the difference in course emphasis. Selection of the items was based on the learning goals for each course. To gauge the affective domain, students in these two courses were given the opportunity to self-assess their mathematical disposition both at the beginning and the end of the course, through an attitude survey.

Details of the assessment program

The attitude survey was administered in both courses. In spite of the level of the material in these courses, some students struggle in these courses and are generally intimidated by mathematics. Through discussions with departments whose students take these courses, faculty who teach these courses, and student evaluations, we felt that an attitude survey might reveal students' beliefs about mathematics and how they approach a mathematics course. An attitude survey was constructed and focused on four areas: confidence, anxiety, persistence, and usefulness. With each of the four areas, a 5-point Likert scale allowed students to choose a descriptor that best described their self-perceived ranking in relation to these four areas. The four constructs along with the extreme descriptors are:

Confidence
    1 = Distinct lack of confidence in learning mathematics
    5 = Confident in one's ability to learn math and perform well on mathematical tasks
Anxiety
    1 = Dread, anxiety and nervousness related to doing mathematics
    5 = Feels at ease in a situation requiring mathematical thinking
Persistence
    1 = Lack of involvement in problem solving; easily gives up
    5 = Keeps persisting in order to complete a mathematical task or problem
Usefulness
    1 = Believes mathematics is not useful now or in the future
    5 = Believes mathematics is currently useful and will be important in future activities and career

The descriptors for "Usefulness" were changed during the spring semester to reflect the use of mathematics in one's personal life as follows:
    1 = Believes mathematics is not useful to me now or in the future
    5 = Believes mathematics is currently useful and will be important in future activities, both personally and professionally

---

[1] www.maa.org/saum/cases/MtMary-A.html

[2] www.maa.org/saum/cases/MtMary-B.html
[3] www.maa.org/saum/cases/MtMary-C.html

This attitude survey has since been modified several times so that the wording is in the active rather than passive voice.[4]

Additionally, students were also asked to write a brief statement to explain their choice on each scale. Research has shown that most people suffer some mathematics anxiety, but "it disables the less powerful—that is, women and minorities—more" (Tobias, 1993, p. 9). The attitude survey, then, was designed to roughly measure students' perceptions about themselves in relation to mathematics. The anticipation is that instructors can assist those with high anxiety and low confidence, once they are identified. Through minimal intervention and encouragement, these perceptions, hopefully, can be altered to generate more success in taking mathematics.

Revisions based on initial discussions

A second round of discussions resulted in a decision to identify and separate the population that did not pass the competency test into three groups. Group one consists of those students who will ultimately take an algebra course. The Developmental Mathematics course will be renamed as Prealgebra, and its curriculum will provide a better bridge to the algebra sequence. Group two are those students who will only need a mathematics course to fulfill the math competency; they will be placed in a course that will be called Quantitative Reasoning. This course will replace the Mathematics for the Liberal Arts course and its curriculum will reflect this emphasis. The third group consists of students who are not ready to take either course. This last group of students needs to obtain basic arithmetic skills before continuing with other mathematics courses. Therefore, a one-credit preparatory workshop was developed. This workshop will serve as a prerequisite for Prealgebra or Quantitative Reasoning. The realignment of courses and curriculum choices were based on our student population and learning goals addressed in the initial phase.

Findings

Students were given the attitude survey at the beginning and end of the semester-long courses. Responses from students who dropped the course during the semester were eliminated from the initial analysis. We also administered the post-course assessment as described earlier in both lower-level courses. During the spring semester, the assessment was administered with and without calculators in the develop-

[4] www.maa.org/saum/cases/MtMary-D.html

mental course. Students often reported that they failed to do well on the competency test due to the lack of the ability to use a calculator. We wished to see if the availability of calculators made a difference in the results.

Results of the attitude survey are shown below:

Category	Fall 2002 Start	Fall 2002 End	Spring 2003 Start	Spring 2003 End
Confidence	3.11	4.04	3.35	3.84
Anxiety	3.09	3.91	3.07	3.53
Persistence	3.63	4.01	3.84	4.12
Usefulness	4.1	4.31	4.08	4.12
Number of students	44	44	40	40

Based on data collected, we are pleased that students are reporting improved results in all categories. However, these results are only self-reported. It was important to also read the rationales provided to gain further insight in relation to these results. In other words, did something in the course impact these changes? Here is a sample of written responses from the end of the semester:

- I think my confidence has improved slightly over the semester; I feel better about coming to class.
- Has improved because I feel I understand things better now. Feel comfortable asking when I don't understand things. I understand it and try to jump right in. Can ask questions without feeling dumb. I am trying more and paying more attention so I can understand and not dread.
- Sometimes I doubt myself on certain problems with homework or tests. Otherwise I feel much better about my ability to do well on my work.
- I know that the things that we are learning in here will be useful one day, so I work hard to understand them and use them.
- After this course I don't even use my calculator as much, so I feel confident.
- If I don't get a problem correct I used to just leave it alone but now I try it again and if I don't understand I come back to it later or I ask for help.
- I still dread math tests/quizzes, but my nervousness has decreased I think.

These comments support the conclusion that the courses positively changed student perceptions about their ability to do and use mathematics. Other comments indicate that these courses do not influence all students in a positive way. One comment, "Sometimes I feel confused or frustrated with my math problems but its math; math isn't supposed to be easy or fun," indicates an unchanged perception of mathematics.

Other comments such as "I'm going to be a history teacher so I might have to add or subtract years!" and "I'm not going to go into a mathematics career," clearly indicate the need to relate the mathematics that students are learning to real-life situations so that students see mathematics as an important part of their liberal arts education.

We also then examined the attitudes of students who dropped the course. The results of the attitude survey of these students are shown below:

Category	Fall 2002	Spring 2003
Confidence	3.38	2
Anxiety	2.79	2
Persistence	3.66	3.83
Usefulness	4.33	4.5
Number of students	12	3

The data was examined to see if students who dropped the course began their mathematics course with less confidence, higher anxiety, and so forth. From the limited number of students who dropped, there was not a clear difference, although we did observe a slightly higher level of anxiety. We will continue to collect and examine such data.

The results of the placement testing and post-course assessment test are reported for each course.[5] Overall, when test items were compared on a one-to-one-basis, we were not extremely encouraged by the results. Although the percentage of students selecting the correct response increased, this increase was minimal on some items. Items that were more computational in nature showed greater gains, but questions that required the application of more critical thinking or problem solving did not show that same level of improvement. Note that in the spring semester, the items for the developmental mathematics course were administered first without a calculator and then with a calculator. The items that require a determination of what to do with numbers (such as using a ratio, or taking two numbers to determine a percentage) did not perform well on the assessment. Overall, there was not a great increase in success with the items when a calculator was permitted. The exceptions were items related to computation and comparison of numbers and calculating unit prices. To our surprise, the question that asked students to determine how many eggs are left over when 115 eggs are put in cartons of 12 eggs, produced poorer results when students actually had access to a calculator. It appears they were unable to interpret the decimal results when dividing 115 by 12. The multiple choice format allows respondents to choose the correct answer often based on number sense, rather than actual computation, but it appears that students often did not use this approach. This finding is disturbing and will be addressed in the future.

Students enrolled in Mathematics for the Liberal Arts performed better on the post-course assessment than those in the developmental course, but these students also scored higher on the initial competency examination, so it is not proper to conclude that the course created these more acceptable results. The item-by-item results can also be found on line.[6] Note that this assessment was done without calculator.

Use of findings

Initial examination of courses led to a decision to revamp the learning goals and curriculum of all remedial mathematics courses and to add a preparatory remedial workshop for students who score low on the competency examination. Additionally, students will be placed in the Quantitative Reasoning course if they do not need the algebra sequence for their major program. These changes were deemed necessary in relation to the needs of our population. The pilot assessment program has given us valuable information and experience in designing a feasible assessment plan. The attitude survey revealed that we need to continue to make efforts to address the affective domain, and to continue to provide a supportive learning environment for these students. We also need to help students develop an appreciation for mathematics in their daily life by emphasizing real-life applications and the connections to their major. The post-course assessment of skills certainly points to the need for students to continue to develop number sense. This needs to be a pervasive theme in their coursework in the mathematics department.

Next Steps and Recommendations

The new structuring of courses will need ongoing assessment to monitor the success of our students and the impact of these changes. We need to assess the effectiveness of the workshop in preparing students for these remedial courses, and the effectiveness of the new Prealgebra in preparing students for the algebra sequence. We will compare those taking the workshop relative to those students who did not take the workshop. We will continue to develop both the attitude survey and skills assessment and expand the assessment program to include a portfolio of open-ended tasks or pro-

---

[5] www.maa.org/saum/cases/MtMary-E.html

[6] www.maa.org/saum/cases/MtMary-E.html

jects to assess the student's ability to solve problems and use real-world data. The portfolio might also include a self-assessment component. We are currently discussing ways to improve or possibly create a new competency exam that will be aligned with the new goals and objectives set forth for these courses.

Since these remedial courses are primarily taught by part time faculty, we have determined it is important to provide these instructors with a rationale and information in relation to the assessment program, the changes in the course descriptions and goals, and research information on helping students deal with mathematics anxiety. We were awarded a grant from the college to hold a day-long professional development opportunity for the part-time faculty before the beginning of the fall 2003 semester; at that time we shared many of the findings and concerns from our assessment plan.

Reference

Tobias, S. (1993). *Overcoming Math Anxiety*. New York: W. W. Norton & Company.

Assessment of Quantitative Reasoning in Applied Psychology

Robert R. Sinclair, PhD,
Dalton Miller-Jones, PhD, and
Jennifer A. Sommers, MA
Department of Psychology
Portland State University
Portland, OR
sinclair@pdx.edu
millerjonesd@pdx.edu
jsommers@pdx.edu

Abstract. This study reports efforts by the Department of Psychology at Portland State University to assess student knowledge and skills in quantitative areas of Psychology. Preliminary research showed that faculty unanimously agree that *research design* (e.g., distinguishing experimental and correlational research designs), *psychological measurement* (e.g., test reliability and validity), and *statistics* (e.g., calculating central tendency indices, interpreting correlation coefficients) are important competencies for undergraduates majoring in psychology. Each of these competency areas is either directly related to or dependent upon students' quantitative literacy. Moreover, at Portland State, math faculty cover several of these concepts in two social-science statistics courses required for psychology majors. This report describes assessment practices and outcomes for quantitative literacy in these areas. The research included psychology students (majors and non-majors) at all levels of the undergraduate program. We found relatively low levels of competence in areas of quantitative research methodology, with psychology majors outperforming non-majors. Senior-level students who completed our intended methodology course sequence reached acceptable mastery levels. However, there appeared to be little advantage for students who had completed the required statistics sequence as compared with those who had not. These results are interpreted within a "continuous improvement" model where, based on these data, adjustments are made to program planning and individual courses that are intentionally designed to impact our learning goals and objectives.

Background

Psychology has a long history of interest in individual assessment across a wide variety of contexts. Consequently, psychologists have played a leading role in much of the scholarly research on learning and behavioral change. However, only within the last decade have psychologists begun to pay serious attention to assessment of learning in undergraduate psychology programs. Perhaps the most tangible demonstration of this interest is the recent release of *Undergraduate Psychology Major Learning Goals and Outcomes: A Report*, a task force study endorsed by the American Psychological Association (Murray, 2002). The study identified learning outcomes for ten educational goals in psychology.[1] Quantitative literacy plays a prominent role in several goals described in the report. For example, the "Research Methods in Psychology" goal explicitly focuses on data analysis and interpretation. Similarly, the "Values in Psychology" goal emphasizes the utility of the scientific method and the value of using empirical evidence to make decisions.

Many psychology programs heavily rely on mathematics departments to provide their statistical training. This reliance most commonly occurs at the undergraduate level, but some graduate level psychology programs also encourage (or require) students to take quantitative methods courses taught by math faculty. In either case, quantitative literacy is a critical component of the psychology major, since advanced courses assume students have a grasp of basic statistical concepts and understand how to apply those concepts to psychology. Thus, math departments often play critical roles in psychology training. Consequently, strong quantitative literacy assessment efforts provide both psychology and math programs with useful data about whether their courses accomplish each program's educational objectives.

In 1998, the Portland State University psychology program responded to a request by the Dean of the College of Arts and Sciences that all departments identify learning goals/objectives their majors should have achieved at graduation. One motivation for this request was the knowledge that the next round of higher education accreditation review would require a focus on authentic indicators of student learning, not just the traditional set of input data (e.g., the number and kinds of classes taught, student enrollments). The university president also designated assessment of student learning as one of three Portland State University presidential initiatives to emphasize the central role of assessment at the university. In response to these challenges, the

---

[1] www.apa.org/ed/monitor/julaug02/psychmajors.html

Psychology Department crafted an assessment vision involving tracking student learning from our initial introductory courses, through our research methods and experimental psychology courses, to our advanced seminars in industrial/organizational, applied developmental, and applied social psychology. This design enables us to determine whether our programs actually affect student learning by tracking changes in students' performance across our curriculum, and using the empirical data generated by our research to guide decisions about curriculum development.

Assessment in Psychology

Our initiative began with a series of workshops in which psychology faculty generated approximately 50 valued learning outcomes. These outcomes were organized into nine broad learning goals that closely resembled the goals suggested by the APA task force report described above. Faculty also indicated which learning outcomes and goals pertained to each of their courses. We used summary ratings (by faculty) of these outcomes and goals to establish assessment priorities. We then organized the learning goals into three categories: Theories and Issues, Application of Psychology, and Psychological Research Methods. Consistent with our description above, the faculty ratings suggested that Research Methods was a high priority topic because mastery of student learning in this area is closely tied to learning about other aspects of psychology.

The broad area of Research Methodology and Statistics consists of four topics:

- Research Design and the Scientific Method,(e.g., use of experimental, observational, questionnaire strategies);
- Psychological Measurement (e.g., reliability and validity in psychological assessment);
- Statistics, and
- Research Ethics.

Quantitative literacy is an essential component of several of these topics. For example, the statistics area presently focuses on three quantitative literacy concepts: central tendency, variation, and association. Upon graduation, we expect students to be able to conduct and present the findings of basic statistical tests in each of these areas as well as to interpret and critique presentations of these tests in published empirical literature. Similarly, we expect students to master basic concepts in research design and psychological measurement. Although research design and psychological measurement are somewhat different than what might traditionally be regarded as quantitative literacy, students use quantitative skills as they learn about these domains. For example, students must understand the concept of correlation to be able to

grasp differences between forms of reliability and validity.

Finally, we note that our decision to focus on quantitative literacy issues also mirrors one of the key ability areas identified by Portland State University's faculty senate for our graduating seniors:

Quantitative Reasoning and Representation — ability to deepen understanding of the value and need for this type of reasoning, the ability to understand the graphical presentation of data, and to transform information into quantitative and graphical representations.

Our Assessment Research

The main purpose of this case study is to describe our preliminary research efforts to assess quantitative literacy and related concepts. This research provides baseline data for future assessment efforts and empirical support for changes to the curriculum. In particular, our data illustrate some of the ways assessment data can be used to document the impact of a program and to pinpoint areas of particular need in curriculum development. We view assessment as fundamentally aimed at demonstrating the effect of the program on student behavior change, typically defined as increased student mastery over learning goals previously identified by the department. Thus, effective programs should demonstrate high levels of overall performance as well as desired patterns of changes in learning over the course of the program.

Our first programmatic assessment efforts involved the development of a 20-item multiple-choice exam covering topics related to research methodology. The test consisted of questions on research design (e.g., distinguishing experimental and correlational research designs), psychological measurement (reliability and validity), and statistics. The statistics section is the most directly related to quantitative literacy and concerns the portion of the curriculum taught in the math department. The statistics questions focused on very basic statistical concepts, such as calculating central tendency (mean, median) and variability (range) measures and interpreting correlation coefficients. Many other relevant concepts were not included (e.g., hypothesis testing). We also asked four quantitatively oriented psychology faculty to rate the difficulty level of the questions, the difficulty of the distracters (i.e., the three incorrect response options for each question) and the level of cognitive difficulty associated with each question. Using these ratings, we sorted the questions into high challenge and low challenge scales. We also sorted the items into three substantive scales corresponding to statistics, research design, and psychological measurement. In each case, the scores were defined as mean proportions of questions successfully answered.

The test was administered during two academic terms to over 800 students taking a wide array of undergraduate psychology classes (from freshman to senior level). We strategically sampled classes so the participants would represent a broad cross section of our students. This strategy enabled us to capture changes in students' mastery of quantitative literacy topics from the time they entered the major (i.e., at the beginning of the courses in our introductory sequence), to their advanced level courses. Moreover, the non-majors taking these courses serve as sort of a quasi control group for examining psychology majors. That is, we would expect to see greater change in majors as compared with non-majors.

Analyses and Findings

We made three types of comparisons. First, we compared the test scores of psychology majors (who are required to complete a sequence of methodology and statistics courses) to non-majors. Second, we examined majors' changes in test performance as they progress through the curriculum. Finally, we investigated differences by class level in quantitative literacy for all students taking psychology courses, including majors, minors, and non-majors, by class level.

Table 1 in the appendix to our report on the SAUM website presents differences between psychology majors and non-majors on each of the test scores.[2] The total scores of 57% for majors and 46% for non-majors represent low levels of quantitative literacy. However, psychology majors out-performed non-majors on all subtest scores by 9-13%, depending on the subtest considered. Interestingly, the highest scores were for the statistics dimension for majors. The statistics score was the only subtest to meet or exceed 70%, which is commonly viewed as "C" level performance in graded classes. The psychology department requires psychology majors to complete two statistics courses, and statistical topics are either explicitly or implicitly covered in several other courses—perhaps to a greater extent than other research methods topics. Thus, our findings most likely reflect the different levels of focus on these topics.

The generally low levels of performance suggest ample need for improvements in our efforts to address quantitative literacy issues. This preliminary finding is consistent with our experience, as well as those of colleagues at other institutions. Undergraduate students often express a great deal of distaste for, or ambivalence toward, topics related to research methods and statistics. For example, psychological measurement was ranked 45th out of 46 on a recent survey

of our students' interests in topics related to psychology. It is important to note that students may not be to blame for these attitudes. Faculty may need to redouble their efforts to teach these concepts in engaging ways. Finally, one piece of good news for psychology majors is that they showed consistently higher levels of performance than non-majors. These differences provide some evidence of beneficial effects of our program for our students as compared with students in other programs.

All psychology majors must complete a core set of curriculum requirements. Many of these requirements concern quantitative literacy issues, including: (a) relatively basic coverage in our two-course required introductory psychology sequence, (b) specific coverage of quantitative literacy issues in statistics courses taught in the math department (but required for psychology majors), and (c) an intense focus on research methodology issues in our upper-division research methods course. This curriculum is founded on the assumption that each of these courses contributes to students' capacity to conduct research, evaluate published studies, and interpret the results of data analyses in applied contexts.

We examine this assumption in Table 2 of the appendix to our report on the SAUM website.[3] This table examines overall test performance for psychology majors broken down by the number of these required courses they have completed. As a whole, the majors scored 57% on the test. These scores were slightly higher for students who had completed the entire sequence (61%) and lower for students who had not completed any of the sequence (52%). Interestingly, there was no difference in overall test performance for students who had the introductory course(s) only and those who had completed the introductory course and a statistics course (both groups obtained an overall score of 56%). This suggests that future attention needs to be given to the extent to which research-focused courses are having their intended effects in our curriculum.

Assessment efforts involve documenting change across an entire educational experience. Moreover, many of the courses that are not explicitly part of our methodology sequence either implicitly or explicitly address methodology issues. Therefore, a second way to explore the effects of the program on quantitative literacy concerns showing performance changes across class levels. Table 3 of the appendix to our report on the SAUM website shows the test scores both for the entire research sample and only for psychology majors. It is important to note that the entire sample data include the majors as well as the non-majors, so these data

[2] www.maa.org/saum/cases/PSU-Psych-A.html

[3] www.maa.org/saum/cases/PSU-Psych-B.html

underestimate the differences between non-majors and majors across program levels.

As this table shows, students in both groups gain in test mastery at each level of the curriculum. The gains are modest, ranging from 5% gains from freshman to senior level for the majors on psychological measurement to 15% gains from freshman to senior level for the entire sample on high challenge items. There are a couple of clear trends worth noting. First, both for majors and the entire sample, there are small improvements in test scores across the curriculum. These findings suggest students are improving their quantitative literacy skills as they progress through the psychology curriculum. Second, psychology majors show higher performance at each class level. Thus, psychology majors have greater mastery of these skills both at entry and upon completion of the program. These findings may be attributable to non-psychology majors being less facile with and/or less interested in methodological issues in psychology. Finally, we note that the test scores were uniformly higher for the statistics subtest than the research design or psychological measurement subtests, particularly for the psychology majors who, by the time they reached senior level classes, reached a marginally acceptable mean of 76% correct on the statistics test. Although there are multiple interpretations of these data, they appear to show that students who have completed more of the psychology curriculum reach higher levels of mastery on quantitative literacy skills.

Insights

Our preliminary analyses indicate a couple of distinct conclusions about quantitative literacy among psychology majors. First, we noted consistent patterns of improvement across levels of the major, suggesting that our current curriculum benefits students. However, the overall levels of performance on these exams are lower than we desire. Thus, it is important to note several reasons why the test scores might be lower than expected in a typical academic examination context. First, students were not informed in advance that the tests would be administered and were not encouraged to specifically prepare for these test questions. Second, student test performance was not linked to their grades in the courses. Students received extra credit for completing the tests regardless of their performance on the exam. Thus, their motivation to perform well was lower than in the typical context of testing for a grade. This means that their scores should not be interpreted in relation to what faculty might expect of students in a normal testing context. On the other hand, most of the questions were of relatively low difficulty levels and did not address sophisticated topics such as hypothesis testing or statistical significance.

Although there are legitimate reasons to expect students' performance to be lower than might be expected on a graded test, we see ample room for improvement in students' performance on future assessments. Therefore, we have engaged in a series of initiatives designed to improve our quantitative literacy training. These initiatives include:

- Developing introductory-level course assignments that actively engage students in quantitative literacy in psychology before they enter statistics courses taught in the math department. In the past, students received relatively little instruction in quantitative literacy in their introductory courses and were expected to learn many statistics topics in math-taught statistics courses in which examples were less clearly tied to psychology. To help address this problem, we have introduced introductory-level assignments that systematically explore research design, psychological measurement, and statistics in hands-on student work. Examples include requiring students to gather their own research data, conduct basic statistical analyses, and present findings in written form. The goal of these assignments is, in part, to help students contextualize the knowledge they receive in their statistics courses and to help them transfer their knowledge from the statistics courses back into the psychology curriculum.

- Expanding our web-based learning resources related to quantitative literacy. These efforts include posting links to existing resources at various web sites and the development of an on-line lab in which introductory psychology students conduct and report the results of a complete research project. The World Wide Web has many examples of useful statistics resources, particularly for psychology courses. We are capitalizing on those resources by locating, gathering, and organizing web material for our students.

- Improving our strategic planning with faculty who teach research methods to develop standard learning goals for research methods courses and other courses focused on quantitative reasoning. This partnership involves efforts to encourage faculty teaching methodology and statistics courses to more actively participate in assessment research design and to draw from the departmental assessment planning as they construct and revise their own courses.

- Experimenting with performance-based grading systems in which students must demonstrate minimum quantitative literacy proficiency levels to receive a B− grade and with "perform to mastery" systems in which students are given multiple opportunities to demonstrate proficiency on the same set of quantitative literacy topics.

Conclusion

Each year, we make small but tangible improvements to the depth and breath of our assessment initiative. Along the way, we have had many opportunities to learn from our mistakes, and even a few opportunities to benefit from our successes. Perhaps the most important thing we have come to appreciate is the importance of, and the challenges with, aligning course content and course assignments with assessment goals. For example, our decision to consciously focus on quantitative literacy required us to add additional course time to that topic and to cut the amount of time devoted to other topics. These decisions can be complex, emotionally arousing, and even adversarial if not handled properly. However, all of the initiatives described above have been implemented to some degree and, in subsequent research we hope to demonstrate improvements in our students' mastery of quantitative literacy.

Reference

Murray, B. (2002). What psych majors need to know. *Monitor on Psychology*, July/August 2002.

Assessing Quantitative Literacy Needs across the University

M. Paul Latiolais, Allan Collins,
Zahra Baloch, David Loewi
Portland State University
Portland, OR
latiolaisp@pdx.edu, collina@pdx.edu,
zsbaloch@yahoo.com, dmloewi@yahoo.com

Abstract. Portland State University is grappling with what "quantitative literacy" means in terms of student outcomes and faculty expectations of students at the departmental level. Pilot surveys of faculty and students were developed to attempt to explore these issues. Initial data from a survey of students with several disciplinary backgrounds was analyzed to develop hypotheses on the mathematical needs of students.

Background and Goals

Portland State University has engaged in several assessment initiatives aimed at determining what graduates need to know and be able to do when they graduate, both departmentally and in general. Mathematical skills have consistently come up as a clear need of students, expressed both by faculty and students alike. What that means as it is translated from discipline to discipline is not yet clear.

The quantitative literacy research team was organized in the Fall of 2001 as part of the University's assessment initiative. The first objective of the quantitative literacy research team was to develop strategies to help departments identify the mathematical (and statistical) needs of their students. The second objective was to articulate the mathematical needs of students in consistent ways to identify the common needs of all students across disciplines.

Description

In an effort to meet our goals, the quantitative literacy research team developed and pilot tested a faculty survey[1] to generate conversations with departments about what was important and to get an initial sense of what students need to know and be able to do in the mathematical realm. We wanted to use quantitative literacy concepts, as opposed to concepts articulated in traditional mathematical terms for two reasons. First, we wanted to have a different conversation about what was important to people and get away from the common response of what math departments are or are not teaching students. Second, we wanted to get a sense of the context in which the students are being asked to apply mathematical tools. This process would help us understand more about what mathematics the students really needed and also point out to the faculty that context mattered. We pilot tested the faculty survey in Biology, Geography, Psychology and University Studies (the core general education unit). We added statistics to our list of quantitative literacy questions, as a result of feedback we received from respondents. We are in the process of continuing the conversation with those departments.

Based on our initial responses to the faculty surveys, we decided to create a student version. The student surveys would ask students what they felt was important and what they thought their skill levels were. Student survey questions[2] were modeled from the quantitative literacy elements in "The Case for Quantitative Literacy" (Steen, 2001, p.8).

---

[1] www.maa.org/saum/cases/PSU-QL-A.html

[2] www.maa.org/saum/cases/PSU-QL-B.html

The Student Survey

In addition to asking students to provide certain demographic information, the survey contained six sections. The first section addressed student's attitudes towards Quantitative Literacy. The next four sections consisted of sub-areas of mathematics identified in Steen that comprise QL: Confidence with Mathematics, Cultural Appreciation of Mathematics, Logical Thinking and Reasoning in Math and Prerequisite Knowledge and Symbol Sense in Math. The last section on Statistics was added based on feedback we received from prior discussions with faculty. Each section asked readers to rate how strongly they either agreed or disagreed with a given statement. The statements were written to reflect the category they were in (i.e., in the statistics section the statements reflected information regarding statistics; e.g., "I am comfortable with statistics").

The Students. The survey was distributed to an introductory psychology course during the 2002 summer session. Forty-seven students completed the survey, 18 men and 29 women. The mean age of the respondents was 28 with a range of 18 to 41 years. Due to the wide age range, three age groups were created: (1) 18–25, (2) 26–35 and (3) 36 and older. The original distribution of majors contained a wide range, so individual majors were grouped in general disciplinary categories. The sample distribution by major is, $n=13$, education; $n=3$, humanities; $n=7$, natural sciences; $n=22$, social science, and $n=2$, undecided.

Insights: What did we Learn?

Quantitative Literacy Section: The results from these items indicated that the majority of students felt they understood

what quantitative literacy meant and believed it to be an important skill because it applies to both their career and daily activities.

Confidence with Mathematics Section: Most students felt confident in their mathematical ability, but felt scared of math, and sought courses that were not heavily loaded with mathematics. Analysis of Variance (ANOVA) results indicated significant gender differences in five of the items in this section (Figure 1). Women felt less comfortable taking math courses, $(p < .052)$, they were more "scared" of math, $(p < .031)$, used mental estimates less often, $(p < .018)$, felt like they did not have a good intuition about the meaning of numbers, $(p < .05)$, and felt less confident about their estimating skills, $(p < .016)$.

Logical Thinking and Reasoning Section: The majority of students felt comfortable in their overall logical and reasoning abilities. ANOVA results (Figure 2) indicated that women questioned numerical information less frequently, $(p < .051)$, and felt less confident in their ability to construct a logical argument, $(p < .04)$, and felt less comfortable reading graphs, $(p < .016)$. It is important to point out that, although women felt less comfortable reading graphs, they did not feel that way about reading maps. This disparity could be a direct result of perceived negative gender stereotypes (Spencer, 2002 & Steele, 1998) on the part of women. Spencer and Steele studied experiences of being in a situation where one faces judgment based on societal stereotypes about one's group, women in our case. That is, the women in our sample may believe that they are not as good at mathematics as men and hence would react more negatively to mathematical terms. The term "graph" likely carries heavy mathematical implications, unlike the word "map."

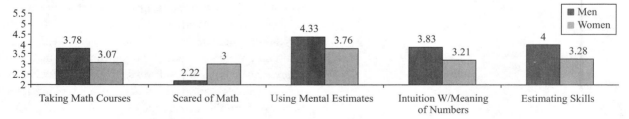

Figure 1. Mean Response by Gender on 5: Confidence w/ Mathematics Items

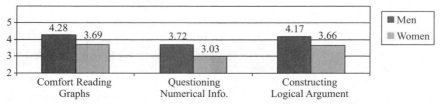

Figure 2. Mean Response by Gender on 3: Logical Thinking & Reasoning Items

Prerequisite Knowledge and Symbol Sense Section: Most students did not enjoy writing proofs and felt uncomfortable interpreting them. In particular, although women felt less comfortable constructing "logical arguments," they were not less comfortable writing proofs. It may mean that the term "proof" is not as understood as the term "logical argument."

Statistics Section: When asked questions about statistics, most students felt comfortable with statistical ideas and in their ability to apply statistical concepts (Figure 3). ANOVA tests revealed that older students felt significantly more comfortable thinking about information in terms of numbers in order to support claims, ($p < .043$).

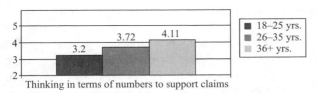

Figure 3. Mean Response by Grouped Age on Item: "Thinking in terms of numbers to support claims."

Cultural Appreciation Section: This section attempted to capture the role mathematics plays in our culture. Approximately 95% of the respondents indicated that math is important and believed that it plays an important role in science and technology. ANOVA results showed that women claimed to be less aware of the origins of mathematics than men ($p < .046$) (Figure 4).

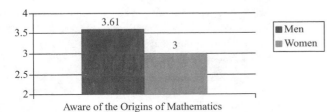

Figure 4. Mean Response by Gender on Item: "Aware of the origins of mathematics."

Successes & Failures

In several of the questions the majority of respondents chose the neutral response. This suggests two possible explanations: 1) that they honestly felt "neutral" with regards to that statement, or 2) they did not understand the statement and that led them to mark the neutral response. In questions where the latter is likely, we may wish to delete the neutral response and reframe the question so that a neutral response is not reasonable.

Respondents overwhelmingly indicated that they understood what quantitative literacy meant. In spite of this finding, it is not possible to gauge whether or not students understand the idea or if they are defining it as mathematics in a more traditional sense. One possibility is to elicit a definition from students at the onset of the survey in order to see how they define it. Our sample was limited and biased towards the social sciences.

Use of the findings

A number of questions have arisen from our results. Why, for instance, are graphs scarier to students than maps? The relatively small size of the sample and its bias towards the social sciences made it difficult to discern information as accurately as we hope to in future studies. Using the survey within departments, we hope to discover:

• Whether the gender differences we report here remain (larger, smaller, nonexistent) in particular disciplines. For example, could it be that women are less confident estimators than men unless they are engineering majors?

• Whether the fear of math is consistent across campus, and hence does the university need to find strategies to address student fear of math courses?

• Whether there are differences in students' understanding of what quantitative literacy means as a function of discipline. For example, do math majors think QL is traditional math skills, political science majors think it means the ability to understand a poll and geology majors think it must include maps?

• Whether differences in comfort level in any of our areas will appear as the disciplines' samples become larger. That is, will differences between humanities students and others become clearer as the number of humanities students surveyed becomes greater than three, or are humanities students really just as confident about logic as science majors?

• Whether there are differences in students' understanding of quantitative literacy as a function of age or work status. For example, do students working full time and only going to school at night feel more comfortable using data to make everyday decisions than traditional, full-time undergraduates?

• If we added a question on whether students voted in a recent election, would this information correlate with their comfort in using poll data, or reading graphs or something else?

Specific Changes to the Survey Instrument. The survey begins with our definition of quantitative literacy. To what extent does that determine the respondents' high comprehension of what quantitative literacy means? We may rewrite the opening section to address possible variations of students' understanding of quantitative literacy and see if this correlates with their majors or other factors.

Based on survey responses, we plan to rearrange some of the questions, delete similar questions whose responses were heavily correlated, and add questions, some from other validated surveys relating to mathematics anxiety. For example, we will change questions in the "Cultural Appreciation" section. We received a high number of correlated responses in this section. Some questions can be eliminated while capturing the same information. We will also re-write questions to make them clearer and more easily comprehensible to students. For example, "*I am aware of the origins of mathematics*" and "*Math plays an important role in technology*" are confusing. We believe students may not fully understand what is meant by those questions. We are also considering changing or deleting the questions relating to the important role math plays in Science and Technology. Students universally agreed that math is important in those areas. If this response is common, then no information is gained by asking the question. But this does raise the question, how does that belief cause students to avoid science and technology courses? Is it because of math anxiety? Or is it due to other factors?

After examining the data from the surveys, we realize we need more information from the respondents related to their math background. We will revise the demographics section of the survey and add questions about the number of math courses students have taken and the grades they received in those courses. These new questions will allow us to compare their math grades with their overall grade point average and to make clearer connections between number of courses and their attitudes towards quantitative literacy skills. A survey designed by the Mathematics and Statistics Department will help us with these questions.

Next steps

One of our goals in this initial phase was to gather data about the survey as an assessment tool. Are the questions readable? Are these questions in line with our assessment goals? Is this the best tool for addressing student attitudes towards Quantitative Literacy? In phase II, we will administer the survey to a group of students who share the same major. We will have data specific to a department. Our goals for phase II are to help departments articulate their quantitative literacy goals for their students and to help them map those skills onto the curriculum.

Unfortunately, due to state budget realities, the funding for the assessment research teams was re-allocated to more direct support for departments to meet the assessment requirements for our accrediting agency. As such, phase II of this work will have to be done piecemeal. Hopefully this work will be a model for the assessments that departments will be doing. Our focus for the future is to work with specific departments and help them identify their students' quantitative literacy needs by administering both the student and faculty survey. The faculty survey examines which quantitative literacy skills faculty members believe are important for their students and where in the departmental curriculum (or outside the departmental curriculum) students should be gaining these skills. Administering both surveys will help departments make progress towards identifying and mapping quantitative literacy skills onto their curriculum.

References

Spencer, Steven J, Steele, Claude M, & Quinn, Diane M. "Stereotype threat and women's math performance." *Readings in the psychology of gender: Exploring our differences and commonalities.* Hunter, Anne E. (Ed); Forden, Carie (Ed). (2002). (pp. 54–68). Needham Heights, MA, US: Allyn & Bacon. xvii, 318pp.

Steele, Claude M. "Stereotyping and its threat are real." *American Psychologist.* Vol 53(6) Jun 1998, 680–681.

Steen, Lynn A., *Mathematics and Democracy: The Case for Quantitative Literacy,* The National Council on Education and the Disciplines, 2001.

Precalculus in Transition: A Preliminary Report

Trisha Bergthold and Ho Kuen Ng,
Department of Mathematics
San Jose State University
San Jose, CA
bergthold@math.sjsu.edu
ng@math.sjsu.edu

Abstract. This report summarizes our initial investigation into low student achievement in our five-unit precalculus course. We investigated issues related to course content, student placement, and student success. As a result, we have streamlined the course content, we are planning to implement a required placement test, and we are planning a 1–2 week preparatory workshop for students whose knowledge and skills appear to be weak. Further study is ongoing.

Background and Goals

San Jose State University is a large metropolitan university in the statewide California State University system. The student body, composed of over 30,000 undergraduate and graduate students in eight colleges is one of the most ethnically diverse in the country, with a large percentage of freshmen being first generation college students.

For several semesters, the San Jose State University Mathematics Department has been concerned about low student achievement in its precalculus course, Math 19. In each of the past several semesters, 40-45% of the 400-500 students who took this five-unit course earned Ds or Fs. All of these students must repeat the course if they wish to take calculus or some other course for which Math 19 is a prerequisite. The financial implications of this outcome to the university are significant: it costs money, time and space to accommodate such a large number of repeat attempts to earn at least a C– in the course.

Three main questions arose. Are the scope and sequence of topics appropriate? Are students being inappropriately placed in this class? What characterizes successful students in this course? Our assessment of factors influencing low student achievement in Math 19 began by addressing these questions.

Description: What Did We Do?

Scope and Sequence of Topics. There are really two issues revolving around the scope and sequence of topics. First, we revisited the topics to be included in the course itself. Second, we wanted to very clearly establish prerequisite knowledge and skills.

Course Topics. To examine the scope and sequence of topics in Math 19, we began by establishing the main purpose of the course. We had always assumed that most students take our precalculus course as preparation for calculus I. To determine whether this was true, we surveyed Math 19 students in Fall 2003 (474 students enrolled, 376 respondents) asking about their intended major, their reason(s) for taking Math 19, and their intention to take calculus. Overwhelmingly, the survey responses indicated that the vast majority of our precalculus students (80%) intended to take calculus I. Of those intending to take calculus, 47% cited some area of engineering as their intended major and 39% cited some area of science as their intended major. (Survey data from Spring 2004 (231 students enrolled, 153 respondents) had similar results.) Based on these results, we believe it's reasonable to focus the topics of Math 19 on preparation for calculus I.

To refocus Math 19 on preparation for calculus I, we streamlined the topics to emphasize depth over breadth. We retained only those topics we felt were absolutely crucial to success in calculus I: functions and their graphs, polynomial and rational functions, exponential and logarithmic functions, trigonometric functions, analytic trigonometry, applications of trigonometry, polar coordinates, analytic geometry, and systems of equations. In several of these broad topics, we eliminated or de-emphasized some elements that seemed to take up extensive class time without contributing substantially to the main goal of preparing students for calculus I. In particular, we eliminated scatter diagrams and data analysis; complex zeros and the fundamental theorem of algebra; simple harmonic motion and damped motion; vectors and operations on vectors; rotation of axes, polar equations, parametric equations of conics; matrices, determinants, and systems of inequalities. In addition, we chose to de-emphasize rational functions, applications of exponential and logarithmic functions, and angular measures in degrees.

By streamlining the topics, we were able to build considerable leeway time into the syllabus. In fact, 30% of the class time is now considered "leeway" and left to the instructors to use as they see fit. This permits them to design in-depth study of topics that are particularly difficult for students. In an end-of-semester survey of spring 2004 Math 19 instructors, we received some indications that they perceive this new approach to be more efficacious than the previous approach. One instructor commented, "The elimination/reduction of some of the topics from the old syllabus (e.g., rational functions, synthetic division, etc.) allowed more time to investigate core topics in a deeper fashion. I would not recommend any additional major changes to the new syllabus."

Prerequisite Knowledge and Skills. To establish the prerequisite knowledge and skills, we gathered information from several sources. First, we looked at the precalculus textbook's (Sulivan's *Precalculus*, 6th ed.) review material contained in the appendices. Second, we looked at the topics covered in the California State University entry-level math requirement[1]. Third, we looked at the topics covered in an online mathematical analysis readiness test produced by the CSU Mathematics Diagnostic Testing Project.[2] These three sources allowed us to create a document listing all prerequisite knowledge and skills that students should have upon entering Math 19, which can be handed out to students on the first day of class.

Independently of our construction of a prerequisite skills list, we asked Math 19 instructors to each provide a list of topics they felt their students should know but did not know upon entering the class. Overwhelmingly, instructors felt that their students' knowledge and skills pertaining to fractions, order of operations, and algebraic expressions were very weak. We intend to use our prerequisite knowledge and skills list along with our instructors' impressions of students' greatest weaknesses to create a one or two week intensive preparatory workshop for Math 19 to be conducted in the week(s) immediately preceding the start of each semester.

Placement Practices. Current placement procedures for calculus I (Math 30, 3 units) require students to achieve a sufficiently high score on a calculus placement exam (CPE).[3] We also have a calculus I with precalculus review course (Math 30P, five units) in which students can enroll by achieving a sufficiently high score on the calculus placement exam or passing precalculus (Math 19) with a C– or better. Students who prefer not to take the CPE are not allowed to take Math 30 or Math 30P; instead, the highest class they are allowed to take is Math 19, assuming they qualify. To qualify to take Math 19, they must satisfy the California State University entry-level math requirement. This can be satisfied with sufficiently high scores on the ACT (23) or SAT (550), or with a sufficiently high grade (C or better) in a transferable college-credit math course taken at a community college, or with a passing score on the CSU entry-level math exam (ELM).

We began to investigate the effectiveness of these placement practices by analyzing Math 19 course grades versus our current entry-level mathematics (ELM) exam scores. These data are summarized in Table 1. The overall percent passing Math 19 in Fall 2003 (66%) was higher than in previous semesters, but it still seems low. Students exempt from the ELM exam (about 50% of the Math 19 enrollment) seemed to do best (72% passed, mean grade of 2.2), which makes sense, since they are likely better prepared than students who are required to take the ELM exam. Students having passed the ELM exam did seem to have a significantly better chance of succeeding in Math 19 (66% passed, mean grade of 2.0) than those who failed the ELM exam (46% passed, mean grade of 1.3). The low grades of stu-

---

[1] The California State University has a system-wide placement testing program in basic mathematics skills that consists of the Entry Level Mathematics (ELM) examination. Further information can be found at www.calstate.edu/AR/FOM.pdf.

[2] See mdtp.ucsd.edu/test/.

[3] SJSU's calculus placement exam is provided by the California State University/University of California Mathematics Diagnostic Testing Project (MDTP). Further information can be obtained at mdtp.ucsd.edu.

	Number of Students	Mean Grade[b] in Math 19	Percent Passing (at least C-)	Correlation with Math 19 Grade
Overall	474	2.0	66	N/A
Exempt from ELM Exam	232	2.2	72	N/A
Took ELM Exam	242	1.9	60	0.29
Passed ELM Exam	168	2.0	66	0.28
Failed ELM Exam[c]	74	1.3	46	-0.04

[a] Grades are reported here as grade points, where an A = 4.0, A– = 3.7, B+ = 3.3, etc.
[b] Mean grades and passing percents exclude grades of W.
[c] Students who failed the ELM exam met the entry-level math requirement by either completing developmental mathematics coursework at SJSU or completing a transferable college-credit math course at a community college with a sufficiently high grade (C or better).

Table 1. Math 19 Grades[a] versus ELM Exam Scores (Fall 2003)

dents who met the entry-level math requirement by other means are a concern. Such developmental mathematics coursework might not be enough to prepare students for Math 19. Perhaps an additional intensive review before the beginning of Math 19 is necessary for these students. Perhaps a better screening criterion or a reorganization of the Math 19 content is needed. We will continue to monitor the effect of our new syllabus.

As a second step in our investigation of placement practices, we analyzed Math 19 course grades versus calculus placement exam (CPE) scores, for those students who took the calculus placement exam (about 12% of the Math 19 enrollment). These data are summarized in Table 2. Students who took CPE before attempting Math 19 did seem to perform significantly better in the course (87% passed, mean grade of 2.7). It's possible that this is due to a biased sample of students taking the CPE. Students who have met the entry-level mathematics requirement may choose to take the calculus placement exam. Given that the exam costs $20, it's reasonable to assume that many of

those students who opt to take the CPE are fairly certain of their calculus readiness, hence are likely to be among the better-prepared students. The sample size of students who took CPE was relatively small, and results might not be too credible. For example, it is surprising that students recommended for Math 30P in fact did worse than those prevented from enrolling in Math 30P. Since there were only four students in the former group, this could just be a statistical anomaly. Further study should be done before attempting to interpret the data.

To gain a different perspective on how well our students were prepared for Math 19, and as an independent check on our placement system, in Fall 2003 we gave our Math 19 students a test during the first week of classes. The test was strictly for diagnostic purposes and did not count towards students' grades.

The test was necessarily short (six multiple choice questions), so as not to demand too much time from instructors or students. Students were asked to compose three functions and identify the graph of the composite function, find the

	Number of Students	Mean Grade in Math 19	% Passing (at least C–)	Correlation with Math 19 Grade
Overall	474	2.0	66	N/A
Did not take CPE	419	1.9	54	N/A
Took CPE	55	2.7	87	0.18
Took CPE, recommended for Math 30P	4	2.6	75	-0.71
Took CPE, recommended for Math 19, prevented from enrolling in Math 30P	32	3.1	94	0.24
Took CPE, recommended for College Algebra and Trigonometry (Math 8), allowed to enroll in Math 19	19	2.2	84	-0.23

Table 2. Math 19 Grades versus CPE Scores (Fall 2003)

1. Let $f(x) = x - \pi/2$, $g(x) = 4x$, and $h(x) = \sin(x)$. Which of the following graphs represents the composition $ghf(x)$?*

2. Let $f(x) = (12x^2 - 7x - 12)/(x - 2)$. Find the zero(s) and vertical asymptote(s).

3. Solve for x: $\log_{10}(x + 1) + \log_{10}(x - 2) = 1$.

4. A guy wire 80.0 feet long is attached to the top of a radio transmission tower, making an angle of $30°$ with the ground. How high is the tower, to the nearest tenth of a foot?

5. Find the vertex of the parabola $y = -4x^2 + 18x - 13$.

6. Solve the system of equations $y = x^2 - 6x + 9$, $y - x = 3$.

   * Each item had five answer choices.

Figure 1. Diagnostic Test Items

zeros and asymptotes of a rational function, solve a logarithmic equation, solve a right triangle problem, find the vertex of a parabola, and solve a system of two equations (one linear, the other quadratic). (Figure 1 contains these diagnostic test items.) For each item, students were given five answer choices, including at least one "main distractor", that is, an answer that would result from an "almost correct" solution attempt. For example, the solution of the logarithmic equation introduces an extraneous answer, so one of the answer choices for this item was $x = -3$ or 4, which could be obtained if a student skipped the final check.

We scored the test two different ways. First, we calculated a raw score: 1 point for each correct answer, 0 points for each incorrect answer. Then we calculated a partial credit score: 1 point for each correct answer, 1/2 point for each item in which a main distractor was selected, 0 points for all other responses.

For the 346 students (out of 474 enrolled) who completed the test[4], the mean score was 1.5 out of 6, with a standard deviation of 1.2. Allowing partial credit for the main distractors, the adjusted mean score was 2.1 out of 6, with a standard deviation of 1.2. We did not expect students to do very well on this test, because some of these topics would be covered in Math 19. We found that the main weakness seemed to be in solving item 3 (solving a logarithmic equation) and item 5 (finding the vertex of a parabola). For item 3, very few students seemed to have any understanding of how to cope with the logarithmic expressions. For item 5, the most popular response was the answer choice indicating the y-intercept of the parabola.

To discover how much these students had learned in Math 19, we gave exactly the same test to students in Math 30/30P (calculus) in Spring 2004, most of whom had taken Math 19 the previous semester. Again, the test was strictly for diagnostic purposes and did not count towards students' grades.

For the 259 students (out of 328 enrolled) who completed the test[5], the raw and adjusted results were increased to 2.3 out of 6 with a standard deviation of 1.2 and 2.9 out of 6 with a standard deviation of 1.1, respectively. The increase was expected, but the magnitude of the increase was somewhat discouraging. Since the topics had just been covered in the previous semester, we had expected the increase to be more pronounced. We are hoping that our revised precalculus syllabus, which emphasizes depth over breadth, will help prepare students better for calculus. Specifically, we found our calculus students to have particular difficulty (still) with item 3, solving a logarithmic equation.

In a survey of Math 19 instructors at the end of the spring 2004 semester, they noted that, while the topic of logarithms was difficult for students, they did not feel that any more time (in the now-streamlined syllabus) was needed on this topic. We intend to repeat the test in our Fall 2004 calculus courses to assess whether the additional depth of study afforded by the streamlined syllabus in Math 19 appears to have an impact on student's knowledge and abilities.

Characteristics of Successful Students. At the end of the diagnostic test given to our calculus students in Spring 2004, we asked them to write some words of advice to Math 19 students on how best to succeed in Math 19. A total of 216 students responded to this open-ended question, providing a total of 348 individual suggestions. Nearly all (333 total) of the suggestions focused on behaviors and perspectives students should adopt and actions students should take to help themselves succeed. We were struck by how positive most of the responses were. Overall, they indicated that these students (most of whom had successfully completed Math 19) did understand why they succeeded or what their mistakes were in precalculus. A summary of the responses is given in Table 3.

Insights: What Did We Learn?

Spring 2004 is the first semester in which we used the streamlined syllabus. Early feedback indicates that the instructors felt the new syllabus was better suited to the needs of the students and the goal of preparing students for calculus. To monitor this in future semesters, we will conduct surveys of precalculus and calculus instructors.

---

[4] Precalculus instructors were strongly encouraged, but not required to participate in the diagnostic test. Out of 15 sections, data were collected from 13 sections. For two of these sections, it was obvious that the instructor had given the diagnostic test as a take-home assignment, rather than as an in-class test. Data from these two sections were thrown out.

---

[5] Out of 10 sections, data were collected from 7 sections.

Numb. of Students	Suggestion
120	Do all the homework
52	Study hard
30	Attend all classes
23	Ask questions/go to office hours or tutoring
18	Pay attention/participate in class
15	Take good notes
14	Keep up
11	Practice
9	Review constantly
9	Read the book
8	Learn algebra/trig/everything
7	Understand concepts; don't just memorize
5	Take Instructor A (a specific instructor at SJSU)
4	Get into a study group

Table 3. Advice on How Best to Succeed in Math 19

Placement remains a difficult issue. There is sentiment in our department to implement a mandatory placement exam for Math 19, but the data from Fall 2003 suggest that success in Math 19 does not depend entirely or even mostly on a single score on a placement exam or competency exam. Data from our diagnostic test showed that students have quite a few weaknesses in their knowledge and skills upon entering Math 19, despite the fact that many of them have taken trigonometry, precalculus, and/or calculus already. At the same time, most of these students intend to take calculus and to major in science or engineering. In fall of 2004, we will implement a course-wide diagnostic exam in Math 19 that is similar to our calculus placement exam (and produced by the same organization). Data from this new diagnostic exam will be much more thorough than that obtained with our 6-item test, and we hope will inform a decision on how to proceed with a mandatory placement exam.

Our investigation of student success has really only just begun. While we now have a sense of what successful students consider to be the keys to their success, we don't have a good sense of what is causing the remaining students to fail. We have much more work to do in this area. In fall 2004, we hope to expand the items on our student survey to include questions about students' outside commitments, their experience with college level expectations, and their initial perceptions of what they think will be the keys to their success in Math 19.

Finally, we are beginning to realize that ongoing assessment is crucial to distinguishing between real problems and anomalies. It is our hope that further assessment efforts will lead to meaningful change for our Math 19 course.

Acknowledgements

This project was partially funded by a San Jose State University junior faculty career development grant.

An Assessment of General Education Mathematics Courses' Contribution to Quantitative Literacy

Aimee J. Ellington
Department of Mathematics
Virginia Commonwealth University
Richmond, VA
ajellington@vcu.edu

Abstract. This study was conducted to determine the role general education mathematics courses play in the development of the quantitative reasoning skills of students enrolled in those courses. The students' post-test results on a series of quantitative literacy questions were compared with pretest baseline results. The assessment revealed that the quantitative reasoning skills of students are improved through participation in general education mathematics courses.

Background and purpose

The twenty first century promises to be an age of information and technology. Much of the information we gather throughout a normal day is quantitative data. Understanding the world and all of its complexities requires a strong sense of numerical data and the quantitative measures used to gather and evaluate numerical information. Quantitative literacy or numeracy, as it is sometimes called, is an essential skill for well-informed citizens. In a typical college or university, at least half of the population consists of students majoring in a field that is not based in science, engineering, or mathematics. As one quantitative literacy expert recently stated, "numeracy, not calculus, is the key to understanding our data-drenched society" [1]. As a result, the quantitative reasoning skills of all students, whether they are working toward a mathematically intensive major or not, must be addressed by institutions of higher education.

Virginia Commonwealth University (VCU) is an urban institution located in Richmond. The diverse student population consists of 25,000 individuals enrolled in 130 graduate and undergraduate degree programs. The State Council of Higher Education for Virginia (SCHEV) is encouraging accountability in Virginia's higher education institutions by publishing a series of Reports on Institutional Effectiveness (ROIE) [2]. These reports are intended to provide educators, policy makers, and prospective students with information about the academic quality of higher education institutions in Virginia. A forthcoming ROIE will focus on the quantitative reasoning skills of graduates of Virginia's public institutions. Based on the SCHEV mandate, the VCU Assistant Provost requested that the Mathematics Department begin the process of assessing the quantitative literacy of VCU graduates.

The goal of the assessment outlined below is to determine the role of VCU general education mathematics courses in improving the quantitative reasoning skills of students who complete the courses. In particular, this project is to determine if VCU is preparing more quantitatively literate graduates through its general education mathematics courses. The assessment focuses on the impact of four entry level mathematics courses: Math 131, Contemporary Mathematics; Math 141, College Algebra; Math 151, Precalculus; and Math 200, Calculus I.

Math 131 is an activity-based course where students are engaged in developing skills needed in real-world situations. Out of the four courses, Math 131 is the class that most directly covers quantitative reasoning topics. On a regular basis, students convert a real-world situation to a mathematical problem, solve the problem and then apply what they learned to the original situation. The students are encouraged to write about mathematics on a weekly basis. Throughout the semester, they participate in hands-on experiments, are engaged in discussions about quantitative topics, and prepare poster presentations on topics with quantitative components.

The three remaining courses are more traditional mathematics courses that do not explicitly focus on quantitative reasoning skills. Math 141 is a large lecture college algebra course designed to strengthen fundamental skills before students take more advanced mathematics courses. Math 151 covers the traditional precalculus topics in small lecture-based class meetings. Math 200 is taught in relatively small classes with a textbook that features both reform and traditional calculus methods.

Method

In Fall 2001, we conducted a search for an instrument appropriate for assessing quantitative reasoning skills. Several instruments designed by faculty at various higher education institutions were evaluated, but none were considered appropriate for the VCU project. As a result, a committee was formed to design a test instrument for the assessment project. Through a series of brainstorming sessions a set of twenty-five multiple choice questions was created. A preliminary test of the questions was conducted with students taking Math 131 and Math 151 final exams in May 2002. Student responses were evaluated to determine whether the questions were appropriate for the assessment and the level of difficulty of the questions. Feedback on the questions and responses was also gathered from colleagues from colleges and universities participating in an assessment workshop sponsored by the Mathematical Association of America. Based on these activities, changes were made to existing questions, some questions were eliminated, and other questions were added yielding a final set of sixteen quantitative literacy questions. Two examples appear in Appendix A; the complete list of questions can be obtained from the author. The multiple-choice questions cover a wide range of quantitative literacy topics. A list of topics and the questions used to assess each topic appear in Table 1 of Appendix B. Many questions in the test bank cover more than one numeracy category.

More than a decade ago, four items on the mathematics placement test were made available for assessment projects. For the current assessment project, four questions from the bank of quantitative reasoning questions were used as the placement test assessment project items. Since there are four versions of the placement test each one contains four different assessment project questions. The same four sets of four questions were included on versions of final exams given at the end of the courses. The types of questions were equally distributed on the placement tests and the final exams.

Test data was gathered over three semesters from Fall 2002 to Fall 2003 from students taking both the placement test and one of the aforementioned courses. No students were excluded from the assessment. Through an extra credit incentive, students were encouraged to answer the questions on the final exams. They were given a bonus of one point on the exam grade for each question answered correctly. Table 2 in Appendix B contains the number of students completing each version of the placement tests and the final exams. The data and results reported below are for all four courses combined, as well as for Math 131 alone.

The data gathered on correct responses to each question on the mathematics placement test was used to generate a baseline percentage. The percentage reflects the proportion of the incoming student population with an understanding of the numeracy concept upon which the question was based. For each question, a statistical comparison was conducted of the proportion of students answering the question correctly on the placement test with the proportion of students who took the placement test answering the question correctly on the final exam. The combined results for all classes as well as the results for Math 131 alone were analyzed. A significant result ($p < .05$) was obtained when the proportion of students answering the question on the final exam correctly was statistically larger than the proportion of students answering the question correctly on the mathematics placement test. This was used to determine that taking a VCU general education mathematics course had resulted in a larger proportion of students understanding the quantitative reasoning concept covered by that question.

Findings

Table 3 of Appendix B contains the combined results for all classes. The placement test column contains the baseline percentage of correct responses to which the final exam percentage of correct responses was statistically compared. The percentages listed in the final exam column that are marked with an asterisk are the percentages that were statistically larger than the corresponding placement test percentage. The percentage of students answering the final exam question correctly was significantly larger than the baseline percentage for all but three questions. Question 2 was a graph interpretation question with a low level of difficulty. Questions 7 and 14 were more difficult questions on proportional reasoning and percent increase.

While numeracy topics are not a direct focus of instruction in Math 141, 151, and 200, it appears that all general education mathematics courses are helping students develop quantitative reasoning skills. Students in these courses are at least indirectly exposed to the topics featured in the assessment questions. Unit analysis is part of the college algebra curriculum. Exponential functions are featured in precalculus. All three courses spend a great deal of time on the analysis of graphs and the use of mathematical functions.

Since Math 131 is the course that specifically covers quantitative reasoning topics, the results for this course were analyzed separately. The percentages appear in Table 4 in Appendix B. There was concern that one aspect of the placement test might result in low baseline percentages. A placement level is generated by the number of correct responses minus one-fourth of the number of incorrect responses. As a result, the instructions state that if a student is uncertain about a response, it is acceptable to not provide an answer to that question. In Fall 2002, over 40% of students placed in Math 131 left questions 4 and 8 blank when they took the placement test. However, this was not the case in Spring and Fall 2003. The low percentages (questions 7 and 10) are not due to a large number of students not answering the questions. They are simply the result of a small number of students choosing the correct response.

The results for Math 131 were mixed. No topic-related patterns could be drawn from the statistically significant results. For example, for the two questions that featured proportional reasoning, the percentage of students answering one question correctly on the final exam was significantly greater than the percentage of students answering the same question correctly on the placement test. The result for the other question was not statistically significant. This was true for all but one of the numeracy constructs that were assessed by two or more questions. Based on the results, the ability to interpret charts and graphs is no different after students complete Math 131.

Use of Findings

The findings for all classes reveal that general education mathematics courses at VCU are playing a role in helping students develop quantitative reasoning skills. The results are also pinpointing aspects of quantitative reasoning that are difficult for students to understand. Determining the relevance of these topics to college algebra, precalculus, and calculus will be the focus of discussion among VCU mathematics faculty. Based on these discussions, we are considering changes in the curricula of these courses. In particular, VCU is planning to teach several small sections (approximately 35 students) of Math 141 with a textbook that highlights problem solving and the use of real-world

applications. The graphing calculator will be a major tool in these sections to allow the course to focus on conceptual understanding with less emphasis on skill development when compared to a traditional college algebra course.

As noted above, for all but three questions in the combined results, the percentage of students getting a question correct on the final exam was significantly higher than the percentage of students getting the same question correct on the placement test. Nevertheless, the final exam percentages for many questions were still low. For the combined results, the percentage of correct responses on final exams was less than 50% for 11 of the 16 questions. The same is true for students who completed Math 131. Therefore, we will be looking at ways to ensure that larger percentages of students complete general education mathematics courses with an understanding of these numeracy topics.

The results for Math 131 in particular do not reflect the level of quantitative reasoning we hope students are developing as they take the course. One possible reason is that many aspects of Math 131 can not be assessed through the set of multiple choice questions that were developed for this project. Traditional instructor-led lectures take place in only one-third of the course time while the rest of the class time is spent with students engaging in group discussions, hands-on activities, and working on long term projects. Written responses to assigned problems are required. In addition to daily assignments, students use their quantitative reasoning skills to write several papers in the course and give a poster presentation at the end of the semester. All student work is graded and none of the assessment instruments feature multiple choice questions. Based on the fact that 69.4% of the students who took the course during the time frame of this project received an A, B, or C as their final grade, the majority of students taking the course are achieving the quantitative reasoning goals established for the course. We will continue to analyze the topics covered and the methods of assessment used in Math 131 to determine whether or not changes are warranted.

Through this assessment, the VCU mathematics faculty has acquired valuable information upon which to base curricular changes. In particular, the students who participated in this assessment had difficulty with the questions involving unit analysis, proportional reasoning, and percent increase or decrease. The VCU mathematics faculty will be looking for ways to improve the instruction in the content areas related to these questions.

Success Factors

Several factors helped make this assessment project a success. First, having several questions on the mathematics placement test available for assessment purposes was essential for gathering baseline data. Second, pilot testing the questions and getting input from colleagues at an assessment workshop helped solidify our bank of questions. The workshop discussions also were beneficial in narrowing the focus of the assessment project. Lastly, the support of VCU faculty was essential to the project. All instructors of general education mathematics courses readily made the assessment questions the last page of their final exams and gave their students extra credit for each question answered cor-

rectly. Without their willingness to participate, this project would not have been possible.

References

1. State Council of Higher Education for Virginia "2002 Reports of Institutional Effectiveness" webpage http://roie.schev.edu/

2. Steen, L. A. "Numeracy: The New Literacy for a Data-Drenched Society." *Educational Leadership*, *57* (2), October 1999, pp. 8–14.

Appendix A. Examples of Questions Used to Assess Quantitative Literacy

Example #1   An "A" tent is one that is open in the front and back and has no floor. For an outdoor project you need an "A" tent that is 10 ft long, 8 ft high and has a 12 ft wide opening. What are the dimensions of the tarp you need to construct the tent?

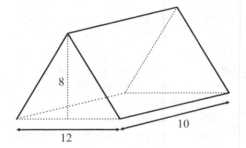

    a. 10 ft × 12 ft   b. 12 ft × 20 ft   c. 10 ft × 20 ft

    d. 10 ft × 24 ft   e. 16 ft × 10 ft

Example #2   The population of a county is 100,000. A power company predicts that the population will increase by 7% per year. The county supervisors predict that the population will increase by 7,500 each year. Which group's prediction method predicts the larger population in 10 years?

    a. supervisors

    b. power company

    c. both predictions are the same after 10 years

    d. There is not enough information provided to answer the question.

Appendix B. Analysis of Quantitative Literacy Assessment

Topic	Question Number
Unit Analysis	1, 7, 9, 13
Interpretation of Charts and Graphs	2, 6, 10
Proportional Reasoning	7, 13
Counting Principles	8
General Percents	3, 6, 11
Percent Increase or Decrease	4, 12, 14
Use of Mathematical Formulas	1, 5, 9, 12, 15, 16
Average	16
Exponential Growth	12

Table 1. Breakdown of Questions by Quantitative Literacy Topic

	All Classes		Math 131	
Test Version	Placement Test	Final Exam	Placement Test	Final Exam
1. Questions 1 – 4	452	489	131	126
2. Questions 5 – 8	551	454	158	127
3. Questions 9 – 12	506	520	140	157
4. Questions 13 – 16	449	495	125	144

Table 2. Number of Student Responses by Question

Question Number	Placement Test	Final Exam	Question Number	Placement Test	Final Exam
1	34.96	45.60*	9	29.11	42.69*
2	84.73	81.80	10	11.07	17.88*
3	25.44	33.95*	11	63.17	75.19*
4	24.78	39.26*	12	22.73	42.88*
5	30.67	42.29*	13	58.48	65.25*
6	38.48	46.26*	14	15.85	18.18
7	13.25	16.52	15	65.70	86.26*
8	24.14	44.49*	16	41.87	63.23*

* Significantly larger percentage, $p < .05$

Table 3. Percentage of Correct Responses for All Four Courses

Question Number	Placement Test	Final Exam	Question Number	Placement Test	Final Exam
1	35.11	31.75	9	26.43	36.31*
2	82.44	78.57	10	8.57	12.10
3	19.85	25.40	11	58.57	76.43*
4	19.08	25.40	12	15.00	38.85*
5	26.58	37.80*	13	43.55	58.33*
6	30.38	35.43	14	16.00	15.97*
7	9.49	11.02	15	53.60	77.08*
8	22.15	37.80*	16	36.00	58.33*

* Significantly larger percentage, $p < .05$

Table 4. Percentage of Correct Responses for Math 131, Contemporary Mathematics

Mathematics-Intensive Programs

Assessment of Mathematics-Intensive Programs

Dick Jardine
Keene State College
Keene, NH
rjardine@keene.edu

Many readers of this volume are mathematicians who do mathematics simply for the sake of the mathematics, just as some people who run enjoy running for the sake of running. But some people run not as an end in itself, but as a means to some other end. Some run to get into shape for another sport, to lose or maintain weight, or for a variety of other reasons. Similarly, mathematics departments offer courses for students who enroll not to learn mathematics as an end, but to acquire the mathematical knowledge and technical skills prerequisite for success in other disciplines. It is far easier to assess how well a person runs than it is to assess how well a mathematics program promotes effective student learning of desired mathematical knowledge and skills. Fortunately, some of our colleagues at institutions across the country have taken up the challenge of assessment in their mathematics-intensive programs.

Mathematics-intensive programs are an area ripe for assessment. Loosely defined, mathematics-intensive programs include the various service courses that mathematicians teach to future engineers, medical specialists, architects, economists, and other professionals, including those courses that serve both mathematics majors as well as students majoring in other disciplines. Since accreditation agencies and university administrators generally focus assessment either on the major or on general education, there has not been as much pressure on mathematics departments to assess mathematics-intensive programs.

Of course, there are specific exceptions to that general statement. For many years, mathematics departments at institutions with engineering schools have had to assess their programs in accordance with the standards established by the Accreditation Board for Engineering and Technology (ABET). Accrediting agencies, and there are many such organizations recognized by the Council for Higher Education Accreditation,[1] are increasingly focusing their evaluations on "learning outcomes" (student work) that are evaluated as evidence that students have learned what departments report they are teaching. The focus of this new type of assessment is not on what mathematics courses students have taken or on what content they have studied, but on what mathematics students demonstrate they can do. Barbara Moskal at the Colorado School of Mines describes an example of a mathematics department with a well-documented, outcomes-based assessment program at an engineering school (p. 149).

Like departments at engineering schools, mathematics departments at institutions with accredited teacher prepara-

---

[1] www.chea.org/Directories/index.asp

tion programs must conduct outcomes-oriented assessment as directed by state education departments and by accrediting agencies such as the National Council for Accreditation of Teacher Education (NCATE). Assessment of mathematical programs that support teacher preparation is addressed elsewhere in this volume, in particular in case studies from Monmouth University (p. 125) and the University of Texas, Brownsville (p. 133).

It is important that mathematics departments take note of and react appropriately to this shift in focus to outcomes-based assessment required by NCATE, ABET, and other professional, state, and regional accrediting agencies. This change in the accreditation process poses a potential threat to mathematics departments that do not keep pace. For example, ABET now requires that students "demonstrate proficiency" in completing mathematical tasks as opposed to just taking mathematics courses. Unless mathematics departments are proactive and conduct assessments of student proficiency in required outcomes, engineering departments could take on the responsibility of teaching and assessing the mathematical content of their programs.

Commonalities

In reviewing case studies in this volume that describe the assessment of mathematics-intensive programs, some commonalities are noticeable. One subtle theme that arises repeatedly is the need to account for the culture of the institution in which the assessment is taking place. Failing to account for the cultural environment can reduce the effectiveness of assessment efforts. Thomas Rishel's report from Qatar (p. 113), for example, illustrates how assessment both revealed and adjusted to cultural issues due not only to nationality differences, but also to the educational background of the students and the perceived need to develop a mathematics program appropriate for the future profession of the students. Assessment can be conducted more readily in an environment in which it is accepted as part of the culture of the institution, as it is at West Point (p. 103), or in relatively cohesive departments such as Keene State College (pp. 157), than it can in departments that are fragmented or not supportive of program assessment.

Computer technology is part of the culture at many institutions, and the use of technology in learning mathematics or in assessing the learning of mathematics is another common theme. Some mathematics departments have the additional service responsibility of introducing students to the use of technology and software such as spreadsheets, computer algebra systems, and/or graphing calculators to solve mathematical problems. Those skills and tools will be used in subsequent courses in other disciplines. To facilitate the use of mathematical skills in courses where they are needed, the new CUPM Curriculum Guide (CUPM, 2004) strongly recommends collaboration between mathematics departments and partner disciplines, including collaborations that ensure the appropriate use of technology across disciplines. This is a significant part of the assessment program at West Point (p. 103) and at Virginia Tech (p. 109).

In the 21st century, computer technology will not only be used for solving mathematical problems but also for the delivery, management, and assessment of mathematics instruction. Course management systems such as Blackboard and WebCT are becoming increasingly sophisticated in their capability to organize, deliver, manage, and assess mathematics courses conducted in the classroom, on-line, or a combination of both. The collaborative use of those tools between disciplinary partners can facilitate the collection and analysis of assessment data. The cost-effectiveness of on-line mathematics instruction which provides, arguably, the equivalent level of mathematical learning is discussed in Virginia Tech's case study (p. 109), a continuation of the assessment described originally by Olin and Scruggs (1999). Electronic portfolios maintained by students are part of the West Point case study, but more sophisticated web-based electronic portfolios are commercially available, and are in use on my campus (Keene State College) by other disciplines.

Using technology effectively is just one example of the important lessons to be learned from the experiences of those who have done assessment in mathematics-intensive programs. The larger critical need is for communication with other disciplines throughout the assessment process. Our colleagues in mathematics' partner disciplines are significantly affected by the learning that occurs in our courses, and it is important that we inform them of, and include them in, our assessment agenda.

Team teaching can be that kind of inclusive interdisciplinary activity, and assessment of such an effort is described in an earlier case study from Jacksonville University (Repsher & Borg, 1999). Conducting the assessment in the courses of other disciplines, as has been done at North Dakota State University (p. 93) can provide "a detailed picture of those quantitative skills needed for upper-division course work in other departments and an assessment of the quantitative capabilities of emerging juniors outside the context of specific mathematics courses." Other activities that promote meaningful communication in interdisciplinary collaboration are addressed in earlier case studies from Oakland University (Chipman, 1999) and the University of Wisconsin at Madison (Martin & Bauman, 1999).

In the latter study, Bill Martin implies that an activity to facilitate interdisciplinary connections is for a member of the mathematics department to participate as a member of the college or university assessment committee. My service on just such a committee helped me make interdisciplinary connections at my college and has ensured the mathematics department is doing assessment work consistent with the work of other departments on campus.

If there is a final theme to address, it is that assessment is a marathon, not a sprint. In the past, accreditation visits and program reviews were periodic events that prompted some last-minute data gathering to support oftentimes vague claims of departmental effectiveness. Motivated by the pressure from institution administrators and accrediting agencies, the experience of many departments is that continuous assessment of specific goals and objectives is becoming part of the department culture, part of the way that departments routinely go about the business of helping students learn mathematics. This is as true in mathematics-intensive programs as it is in the other programs for which mathematics departments are responsible.

References

Chipman, J. Curtis. "'Let them know what you're up to, listen to what they say:' Effective Relations Between a Department of Mathematical Sciences and the Rest of the Institution." In *Assessment Practices in Undergraduate Mathematics*, Bonnie Gold, et al., eds. Washington DC: MAA, 1999, pp. 205-208.

Committee on the Undergraduate Program in Mathematics. *CUPM Curriculum Guide 2004*. Wasington, DC: MAA, 2004, p. 32.

Martin, William O. and Steven F. Bauman. "Have Our Students with Other Majors Learned the Skills They Need?" In *Assessment Practices in Undergraduate Mathematics*, Bonnie Gold, et al., eds. Washington DC: MAA, 1999, pp. 209–212.

Olin, Robert and Lin Scruggs. "A Comprehensive, Proactive Assessment Program." In *Assessment Practices in Under-graduate Mathematics*, Bonnie Gold, et al., eds. Washington DC: MAA, 1999, pp. 224–228.

Repsher, Marilyn L. and J. Rody Borg. "TEAM Teaching Experience in Mathematics/Economics." In *Assessment Practices in Undergraduate Mathematics*, Bonnie Gold, et al., eds. Washington DC: MAA, 1999, pp. 213–215.

Developing a Departmental Assessment Program

William O. Martin and Dogan Çömez,
Department of Mathematics
North Dakota State University
Fargo, ND
william.martin@ndsu.nodak.edu
dogan.comez@ndsu.nodak.edu

Abstract. This article describes the process used to develop assessment in the mathematics programs at the North Dakota State University (NDSU). The Mathematics Department has developed a comprehensive assessment process that examines student learning in (a) services courses, (b) the major, and (c) the masters and PhD program. The most ambitious component, established with external funding, examines the introductory mathematics courses in conjunction with the NDSU general education program. Assessment of the undergraduate and graduate programs involves many of the Department's faculty. All components of the project are designed to minimize extra demands on participants, to provide useful information for participants as well as the Mathematics Department and University, and to focus on assessment as an integrated part of departmental activities rather than an "add on" activity done primarily for external purposes.

Context and setting

North Dakota State University is a land grant, Doctoral I research university, and is the top institution in the state for graduating agriculture, engineering, mathematics and science students with baccalaureate through doctorate degrees. The number of undergraduate students (Fall 2002) is 9,874; and the number of graduate students is 1272. The average ACT composite score of all entering students (Fall 1997) is 23.1 (the national average is 21.0). The student to teacher average ratio is 19 to 1. Most of the classes specifically relating to the majors typically have fewer than 25 students, and mostly research faculty with terminal degrees teach those courses. The normal teaching load for research faculty is four courses per year.

The Department of Mathematics at NDSU offers BS (mathematics and secondary mathematics education), MA, and PhD degrees. The Department also has a major service role for other science and mathematics-intensive programs in the institution, particularly in the Colleges of Science and Mathematics, Engineering, Business Administration, and Pharmacy. The Department offers a broad and balanced curriculum of courses with 15 tenure-track faculty and about 10 lecturers (Computer Science and Statistics are separate departments). In Fall 2002 there were 38 mathematics majors in sophomore-senior standing among 83 undergraduate majors. Many talented students participate in the EPSCoR-AURA program; mathematics faculty members frequently supervise the undergraduate research projects of talented mathematics students. The undergraduate mathematics major's degree program culminates with a capstone course, usually completed during the senior year. The Department, as the largest service department on the campus, enrolls 300–400 students each in calculus I and calculus II every semester (taught in a large lecture and recitation format) and 150–300 students per semester in each of calculus III and differential equations (taught in classes of about 35 students). The Department provides free tutoring services for all 100–300 level mathematics courses, staffed mostly by graduate students and talented undergraduate mathematics and mathematics education majors.

Project goals and program description

Our goal is to develop and conduct a comprehensive assessment program to monitor the impact of all of our instruction on student learning of mathematics. We focus on three components of our instructional role: (a) Service courses through the first two undergraduate years, (b) the undergraduate program for mathematics majors, and (c) the grad-

Mission Statement. The mission of the Department of Mathematics is teaching, research and other scholarly activities in the discipline; providing quality education to our BS, MS and PhD students and post doctoral associates; and influencing the mathematical climate of the region positively. The Department strives for excellence in teaching its majors and service courses, while providing stimulating and informative courses. The Department's research activities include pure and applied mathematics.

Program Objectives (Bachelors program):

1. Students will be able to analyze problems and formulate appropriate mathematical models.
2. Students will understand mathematical techniques and how they apply.
3. Students will recognize phenomena and be able to abstract, generalize, and specialize these patterns in order to analyze them mathematically.
4. Students will be able to express themselves in writing and orally in an articulate, sound and wellorganized fashion.

(Objectives for other programs are in Appendix A.)

Figure 1. Mission and Objective

uate program. The assessment program is designed to involve many departmental faculty in our activities and to coordinate our departmental efforts with the work of the University Assessment Committee.

Development of the program. Two components of our departmental assessment activities have been developed separately: (a) a campus-wide quantitative assessment project focusing on first- and second-year service courses through multi-variable calculus and differential equations and (b) departmental assessment of our undergraduate major and graduate programs. The campus-wide quantitative assessment project uses a model first developed by Martin and Bauman at the University of Wisconsin-Madison (Bauman and Martin, 1995; Martin, 1996) that originally was funded at NDSU by the Office of Assessment and Institutional Research. A recent, more extensive implementation occurred with support from the Bush Foundation of Minneapolis.

The departmental degree program assessment activities were developed to make use of existing instructional activities, reducing financial costs and time demands on faculty. Data is obtained from specific courses required of all undergraduate students, graduate program written and oral examinations, and advisor reports. Additionally, the Department has developed and begun to implement a peer review of teaching program, which will provide additional information about instruction and student learning.

Departmental service role assessment. The most ambitious component of our assessment activities is the quantitative

assessment project. Briefly, the purpose of the project is to gather information about (a) quantitative skills used in specific beginning upper-division courses and (b) the extent to which students can show these important skills at the start of the semester. Instructors play a key role in helping to design free-response tests reflecting capabilities expected of students from the first week and essential for success in the course. Two important characteristics of this form of assessment are (a) direct faculty involvement and (b) close ties to student goals and backgrounds. We have found that the reflection, contacts and dialogs promoted by this form of assessment are at least as important as the test results.

The process begins with the selection of beginning upper-division courses across the campus. These courses are selected either (a) by the Department Assessment Committee or (b) by the instructors themselves. Course instructors, selected from a range of departments, identify the specific quantitative skills their students need. The students are then given a test at the start of the semester designed to determine whether they have these skills. The tests, given early in the term, assess the extent to which students possess those quantitative skills that their instructors (a) identify as essential for survival in the course, (b) expect students to have from the first day of class, and (c) will not cover during the course. The tests are intended to be neither "wish lists" nor comprehensive examinations of the content of prerequisite mathematics courses.

A sample report for Mathematics 265 (University Calculus III, Spring 2002) is available as an appendix to the NDSU Case Study on the SAUM web site.[1] This report was provided to the course instructors, the Department of Mathematics, and the University Assessment Committee. The report includes a copy of the two test versions that were used. In each test we have reported *success rates* for the students who took the test (proportions who successfully answered each question), reported by problem. The report also provides (a) information about the performance of students on each test version, (b) a ranking of problems by their success rates, and (c) information about the grades students earned in previous mathematics and statistics courses.

Corrected test papers are returned to students, along with solutions and specific references for remediation, within one week. Instructors receive information about the students' test performance a few days later. Thus, early in the semester both students and instructors possess useful information about instructor expectations, student capabilities, and the need for any corrective action. We do not prescribe any specific action in relation to the test results, leaving

---

[1] www.maa.org/saum/cases/NDSU-A.html

those interpretations and decisions to the course instructor and students. We do indicate where each type of problem is covered in textbooks used in NDSU mathematics courses so that instructors and students can review the material, if necessary.

We have developed a reliable grading system that allows mathematics graduate students, with limited training, quickly to record information about the students' work and their degree of success on each problem. The coding system provides detailed data for later analysis while allowing the quick return of corrected papers to the students. The sample report for Mathematics 265 cited above includes summary comments about students' performance on the tests.

Information of two kinds is generated by this assessment process: (a) a detailed picture of those quantitative skills needed for upper-division course work in other departments and (b) an assessment of the quantitative capabilities of emerging juniors outside the context of specific mathematics courses. The first comes from personal contacts with faculty members as we design the test and interpret the results; the second is provided by analysis of students' performance on the assessment project tests and their quantitative backgrounds as shown by university records.

Mathematics degree programs assessment. We have also developed a process for assessing learning in the Department's three degree programs: Bachelors, Masters, and Doctoral. Because we have extensive contact with our majors and graduate students over more extended periods than students in service courses, a priority was to make better use of existing data rather than developing new, specialized assessment instruments. Faculty members reviewed the Department's instructional objectives, which had been prepared as part of early assessment activities for the university, and identified existing opportunities to assess learning in relation to these stated objectives. We were able to locate evidence related to all objectives. The evidence was obtained from three main sources: (a) The undergraduate introductory proof course (Math 270, sophomore level) and our capstone course (Math 490, senior level); (b) Graduate qualifying and final examinations; and (c) Graduate student advisors. We developed forms to be completed by faculty members (a) teaching targeted courses, (b) preparing and grading departmental examinations, and (c) advising graduate students. Appendix B contains a sample rating form for the Senior Seminar; forms for other courses and other degree programs can be found in the appendix to the NDSU Case Study on the SAUM web site.[2]

---

[2] www.maa.org/saum/cases/NDSU-B.html

Department Instructional Objectives. The Department had previously adopted a list of objectives for student learning in its three degree programs (see Appendix A). As noted above, we designed rating forms that list objectives that might be assessed through observations in a particular context (for example, the masters comprehensive exam or the capstone course.) Faculty are asked to rate students as fail, pass, or high pass on each outcome. They are then asked to provide descriptive comments about student performance as shown by this assessment or activity to provide evidence that supports their evaluations and to expand on the ratings. These forms are available for faculty members to complete while they conduct the targeted activities. Faculty ratings and comments are based on the standard tools of measurement used to assess and evaluate the student performance in a class, such as classroom tests, quizzes, written assignments, and group work reports. The Department has course descriptions (called TACOs for *Time Autonomous Course Outlines*) for instructors in all undergraduate courses and uses common exams and grading in most introductory courses. These are designed to help ensure a degree of uniformity for sections taught by different instructors and from semester to semester.

Completed forms are returned to the Department Assessment Committee, which analyzes results and prepares a summary report to the Chair, Graduate and Undergraduate Program Directors, and the Department. This process has ensured that a large majority of our Department's faculty are involved in assessment activities each year. At the same time, the extra demands made on individuals by assessment is minimized—most faculty are only asked to provide information they obtained for other reasons and to review and react to the summary assessment report. This is a welcome change for the Chair, in particular, who formerly took responsibility mostly alone for preparing the annual assessment report for the University Assessment Committee and university administration.

Implementation

The assessment program implementation is being done in an ongoing fashion while focusing on one or more courses each year, and continuing the data gathering in the courses whose assessment has begun earlier. To illustrate our implementation process we provide the assessment activities for the academic year 2002–2003.

Aspect of program to be assessed. We chose to focus this year on the three-semester engineering-calculus sequence, introductory linear algebra, and differential equations. The guiding question for our work was "Do students develop the quantitative skills they need for success in later studies in their chosen field?" To respond to this question we investigated three things:

1. What are the existing goals for our introductory service courses?

2. What are the quantitative expectations of our clients (for example, later math courses, engineering programs, physical science programs)?

3. To what extent do students meet the expectations of our clients?

Status of learning goals for this subprogram. We have two kinds of goals for this program. The Department has an explicit objectives statement that covers the undergraduate program, including these courses. These objectives were displayed earlier in Figure 1. This project additionally identifies implicit objectives for the introductory sequence of service courses. Part of the data analysis includes a review of the items that appear on tests. This analysis identifies implicit goals and objectives for the service program. An important part of the project is for the Mathematics Department to review and respond to the findings, including these implicit goals. This took place at assessment committee and departmental meetings during April and May.

Activities during 2002–03. Following the guidelines we set for this year's assessment program, we completed the following activities:

1 *Quantitative Assessment of general education and service courses.* This is the continuation of the assessment process we started seven years earlier and is an ongoing process for the regular calculus sequence; and initiation of the assessment process for focus courses for this year (the three-semester engineering-calculus sequence, introductory linear algebra and differential equations). This part of the program implementation involved more assessment and reporting than analysis and response, particularly for the new courses.

2. *Undergraduate majors.* We had faculty members rate student performance in the introductory proof course and in the senior seminar capstone course.

3. *Graduate students.* Faculty members and advisors rated student performance on exams and progress toward their degree using forms such as the one in Appendix B.

4. *Increased involvement of faculty.* We have wanted to increase faculty involvement in the assessment program for many years. It seemed that having the same small group of faculty conducting the assessment activities did not promote wider faculty involvement, since most assumed the people who had done it before would continue to take care of the work. Working with the Department administration, we adopted a new strategy to increase faculty involvement: Each year a new group of 4–5 faculty (which includes at most two faculty from the previous year) would conduct assessment activities. This strategy worked well. The new members of this year's assessment committee took ownership of the program, carrying the bulk of the activities, but they were not intimidated by the task since they had a good model to use as a template for their activities and reports and experienced faculty members to provide guidance. Formation of the committee for the next year's assessment activities has been significantly easier since more faculty were willing to participate, recognizing that the task did not impose onerous expectations for additional work.

5. *Peer review of teaching.* Several faculty developed a proposal for a departmental peer review of teaching program to complement the limited information provided by student course evaluations. The committee that developed this program began their planning in Fall 2001. The program was adopted by the Department in Fall 2002 and has been piloted by four pairs of faculty or lecturers during 2002–3. Details appear in Appendix C.

6. *Connections to University Assessment Committee (UAC) activities.* One Department member, Bill Martin, has been actively involved in NDSU assessment activities as a member of the UAC steering committee, the University Senate Executive Committee, and the Senate Peer Review of Teaching Board. This institutional involvement has contributed to the integration of Department assessment activities with the assessment work being conducted at NDSU. Consequently, activities conducted in the Mathematics Department have helped to shape the assessment strategies adopted at the university level.

Insights and Lessons Learned

Findings and success factors. The process we have developed takes an ongoing, integrated approach that seeks to embed assessment activities in our instruction. We believe the process provides useful insights to the learning that takes place in our programs. To illustrate the sort of information we obtain, a recent summary report described findings of the annual quantitative assessment project, that focuses on service courses, in this way:

The tests of greatest interest to the Department of Mathematics were given in Calculus III (235 students, four instructors), Calculus III with vector analysis (47 students, one instructor), and Differential Equations (264 students, five instructors). These courses include many students who are majoring in technical programs across the campus, including physical sciences, mathematics, and engineering. All require students to have successfully completed the first year regular calculus sequence. As noted earlier, a sample course report giving detailed information about the outcomes is included as an appendix to the NDSU Case Study on the SAUM website. Faculty members discussed reports of the Fall 2001 tests during a December faculty meeting. The discussions ranged over the nature of the assessment program (for example, whether the tests were appropriate) and the success rates. While faculty members expressed a range of opinions, they agreed that the program was potentially very useful and should continue. These initial results did not lead to specific proposals for course changes this year.

Individual faculty who taught the courses in which assessments were given were asked for their reactions to the test results. The tests revealed areas of strength in student performance along with weaknesses that concern faculty. These patterns were reflected both in the comments at the meeting and in written responses to the reports. There was agreement by many that the information was useful as an indicator of program strengths and weaknesses. More specific information about success rate patterns and their perceived significance is provided in the reports themselves.

So far, our assessment findings have not led to major changes in courses or programs at NDSU. A current focus of our work is on making better use of the information obtained from assessment activities. We plan to have a more extensive review and discussion of findings by departmental faculty, now that we have data from several years. The purpose of the discussion is to address several questions:

1. What do the findings show about student learning and retention from our courses?
2. What might account for these patterns? In particular, why do students seem to have specific difficulties?
3. What could and should the Department do to address areas of weakness?
4. Are we satisfied with the Department's stated goals and our assessment procedures, having attempted to assess student achievement in relation to the stated goals for several years?

While the focus of each test is on a particular course, we are able to gain a broader perspective on faculty expectations and student achievement by pooling results from different assessments and over several years. Figure 2 illustrates the patterns that can be discerned in the results. The table also summarizes some generalizations we can make based on tests administered by the project. We have found three levels of mathematics requirements or expectations in courses across the campus. Within each level, patterns of students' success rates have become apparent over the years.

The course level is based on mathematics prerequisites. For example, Level 2 courses require just one semester of calculus (examples include Finance and Agricultural Economics courses). The success rates range from *High* (where more than two-thirds of the tested students in a class are successful) down to *Low* (when under one-third of the students are able to solve a problem correctly). Each cell reports a general trend we have observed. For example, typically any calculus problem administered to students in a Level 2 course will have a low success rate. The cell also

	Level 1 (no math or stat prerequisites)	Level 2 (require 1 semester of calculus)	Level 3 (expect 3 semesters of calculus)
High success	Basic arithmetic, statistics and conversions (computational) *Example*: temperature conversion	No common items for all subjects fit here; basic statistics is an example *Example*: change in mean	Most precalculus, use calculus formulas and techniques (e.g., differentiate) *Example*: evaluate integral
Mixed success	No common types across most courses at this level *Example*: compare proportions	Precalculus material, such as solving 2×2 systems or reading values off a graph *Example*: profit function	Concepts from calculus *Example*: estimate a derivative or integral from graph
Low success	Extract information from tables and graphs *Example*: 2×2 cross tabulation table	Nearly all calculus material *Example*: estimate derivative at point	Complex numbers, ODE's, series, and more complex word problems (e.g., optimization) *Example*: minimize can's surface area

Figure 2 . Patterns of Student Results

mentions a specific problem to illustrate the trend. The example problem for typically low success rates in a Level 2 course is asking students to estimate the value of a derivative at a point given a graph of the function. The most important characteristic of this table is that it illustrates how the use of tests that are custom-designed for particular courses can still provide detailed and useful information about mathematics achievement on a much broader scale at the institution.

The third appendix in the NDSU Case Study on the SAUM web site displays a more complex table that illustrates how even more detailed information can be extracted from a large number of tests administered across many departments and years. The table illustrates that not only success rates on particular problem types, but even the distribution of types of problems can be analyzed to help identify how mathematics is used across the campus in different programs. This table compares the nature of tests and patterns of success rates in mathematics, engineering, and physical science courses, all of which require the full threesemester normal introductory calculus sequence.

The table is based on 240 individual problem success rates (PSR-success rates for each time a problem was used on a test). The three groups of courses were:

(a) Mathematics (four distinct courses, including a differential equations course that was tested in successive semesters; with 58 PSR);

(b) Physical Sciences (five distinct courses, including a two-course atmospheric science sequence with retested students in successive semesters; 68 PSR); and

(c) Engineering (six distinct courses, two of which—electrical and mechanical engineering—were tested in successive semesters; 114 PSR).

The table is relevant to this Case Study not so much for detailed analysis of its content but to illustrate the detailed information that can be provided by this assessment process.

For example, the table illustrates quite different patterns of mathematics usage across the three disciplinary areas: Mathematics courses emphasized non-calculus material (60% of the problems that appeared on tests in those courses), science courses drew most heavily on differential calculus material (56% of problems), while engineering courses had a more balanced use of problems from across all the introductory areas (22% non-calculus, 31% differential calculus, 16% integral calculus, 26% differential equations, and 5% probability and statistics). Much more detailed information is included about specific types of problems and typical success rates. For example, the first entry for mathematics is "Graph Interpretation" problems which

appeared on two different tests in one math course. These problems represented 3% of all problems that appeared on math course tests, and the median success rate across all problems of this type that were administered in a math course fell in the second quartile representing 25–50% for students taking those tests.

Dissemination of Findings. Our assessment findings have been shared with four distinct groups: (a) Mathematics faculty at NDSU, (b) NDSU departments who depend on mathematics, (c) other NDSU faculty interested in departmental assessment, and (d) mathematics faculty from other institutions involved in the MAA Assessment Project SAUM. The first two groups are most interested in student performance and its implications for their courses and programs. The second pair are interested in the assessment methods employed by our project.

A goal of our work, both in the design of assessment activities and the strategies used to involve faculty and disseminate results, has been to only do things that have value for participants. For example, when we ask students to take tests, we want it to have personal value for them at that time rather than just appealing for their participation for the good of the department or institution. Similarly, when we ask faculty to conduct an assessment in their class or to review reports, they should feel they have gained valuable insights as a result of their work rather than submitting a report because it is required for some external purpose.

Next steps and recommendations

Some of our work requires the assistance of a graduate student to help with test administration and data analysis and some financial support for duplication and test scoring. We have found support for this work through external grants and are working to institutionalize this support as a part of the University's institutional assessment and accreditation activities. The work is valued at the institutional level because the extensive service role played by mathematics is well recognized. Consequently, we expect to receive some level of institutional support for our general education assessment activities, the ones that require the most extra work to conduct and analyze.

We recognize that we have to date had more success gathering and disseminating assessment data than getting faculty to study and respond to the findings. This partly reflects the natural inclination of faculty to focus on their own courses than on the broader picture of how programs are working to develop student learning. We plan to concentrate our efforts now on ensuring that assessment findings are regularly reported and discussed by faculty, both in participating departments and in the Mathematics Department. We believe that regular conversations about the patterns of results will lead to the formulation and implementation of responses to shortcomings revealed by assessment activities. Our approach reflects the belief that faculty are in the best position to respond to findings and that our most important role is in providing accurate information about student achievement. Consequently, our reports focus on providing descriptive statements about student performance, rather than making detailed recommendations for changes in courses and instruction.

We also believe that widespread faculty involvement in assessment activities is a necessary condition for an effective assessment program. Our strategy has been to adopt a non-judgmental approach that seeks to minimize special effort required of participants and to ensure that participants clearly see that they stand to benefit from the activities in which they are involved. Our efforts to increase departmental and university faculty involvement and impact will continue. The strategies initiated during the last academic year seem to work. The Department's assessment committee will continue to work with UAC and General Education Committee to increase the impact of the departmental assessment activities to a broader audience.

References

Bauman, S. F., & Martin, W. O. (May 1995). "Assessing the quantitative skills of college juniors." *The College Mathematics Journal*, 26(3), 214–220.

Martin, W. O. (1996). "Assessment of students' quantitative needs and proficiencies." In T. W. Banta, J. P. Lund, K. E. Black, & F. W. Oblander (Eds.), *Assessment in Practice: Putting Principles to Work on College Campuses*. San Francisco: JosseyBass.

Appendix A. Department Mission Statement and Program Objectives

Mission Statement

The mission of the Department of Mathematics is teaching, research and other scholarly activities in the discipline; providing quality education to our BS, MS and PhD students and post doctoral associates; and influencing the mathematical climate of the region positively. The Department strives for excellence in teaching its majors and service courses, while providing stimulating and informative courses. The Department's research activities include pure and applied mathematics.

Program Objectives

A. Bachelors program

1. Students will be able to analyze problems and formulate appropriate mathematical models.
2. Students will understand mathematical techniques and how they apply.
3. Students will recognize phenomena and be able to abstract, generalize, and specialize these patterns in order to analyze them mathematically.
4. Students will be able to express themselves in writing and orally in an articulate, sound and wellorganized fashion.

B. Masters program

1. Students will have experienced both breadth and depth in the study of advanced mathematics so that they: (a) can recognize and create good mathematical arguments, (b) have knowledge of fundamental topics in both classical and modern mathematics, (c) can create and pursue new ideas and application in and of mathematics.
2. Students will have experience as a teaching assistant with classroom experience or as a research assistant.

C. Doctoral program

1. Students will have experienced both breadth and depth in the study of advanced mathematics so that they: (a) can recognize and create good mathematical arguments, (b) have knowledge of fundamental topics in both classical and modern mathematics, (c) can create and pursue new ideas and application in and of mathematics.
2. Students will have exposure to and experience with current research.
3. Students will develop ability to understand and create new mathematical ideas and applications.
4. Students will have experience as a teaching assistant with classroom experience or as a research assistant.

Appendix B. Sample Rating Forms*

Senior Seminar Rating Form — NDSU Department of Mathematics		
Based on the performance of the _____ students who participated in the Senior Seminar during the _____ semester, I am able to make the following observations about achievement of intended student outcomes based on the objectives listed in the Chart for the Department of Mathematics Bachelors Degree Program. Examiner:_____ Date:_____		
Outcome	**Rating of student performance on this outcome (give number of papers or candidates rated at each level for each outcome)**	**Descriptive comments about student performance shown by this assessment instrument (attach additional pages if more space is required)**
1. Students will be able to analyze problems and formulate appropriate mathematical models.	High Pass _____ Pass _____ Fail _____	
2. Students will understand mathematical techniques and how they apply.	High Pass _____ Pass _____ Fail _____	
3. Students will recognize phenomena and be able to abstract, generalize, and specialize these patterns in order to analyze them mathematically.	High Pass _____ Pass _____ Fail _____	
4. Students will be able to express themselves in writing and orally in an articulate, sound and well-organized fashion.	High Pass _____ Pass _____ Fail _____	

---

* Forms for other courses and other degree programs can be found on the NDSU Case Study on the SAUM web site at www.maa.org/saum/cases/NDSU-C.html.

Appendix C. Mathematics Department Peer Review of Teaching Program

Peer Evaluation of Teaching Proposal

The Department of Mathematics believes that the purpose of peer evaluation is to help faculty recognize and document both strengths and weaknesses in their teaching. The word "peer" means that this activity should involve reciprocal observation and discussion of teaching and learning by small groups of 2–3 faculty who exchange visits in each other's classes. The committee believes that the members of the department have all the qualifications necessary to make this process reach its intended goal. The committee proposes that:

1. Tenure track faculty be reviewed at least once each year; Tenured associate professors be reviewed at least once every other year; Tenured full professors be reviewed at least once every three years.

2. The process begin with the identification of the faculty to be evaluated by the chair. Then the faculty member identifies his/her teaching goals and strategies (in writing). These objectives are discussed with a peer colleague or colleagues, with a view to developing evidence that supports the individual's claims. This evidence could come from classroom observations, student evaluations, and review of written course materials, such as tests and assignments. It should include multiple sources (i.e., not a single classroom observation). After reviewing this evidence, the group prepares a report that describes the activities and the extent to which the evidence supports the original claims. The report should include plans for future teaching strategies, including possible changes or enhancements that the faculty member plans to try.

3. A team of 2–3 faculty members will complete the work described in (2) for each member of the team. This helps to ensure that peer evaluation does not become a one way process that involves one person observing and evaluating another primarily for external purposes. Instead, the process is designed primarily to increase collegiality and reflective practice within the department, while providing documentary evidence of the regular review of teaching that can be used for external purposes (such as annual reviews, PT&E).

4. Observers of a faculty member should include at least one member of the department PT&E committee.

5. The observation process should always include a Pre-Observation Conference between the observee and observer to discuss the objectives of the class to be observed and other relevant issues (see Peer Review Observation Instrument). Following the in-class observation, a Post-Observation Conference must also be held to discuss the observations as documented by the Peer Review Observation Instrument.

Assessing the Use of Technology and Using Technology to Assess

Alex Heidenberg and Michael Huber
Department of Mathematical Sciences
United States Military Academy
West Point, NY
aa5178@usma.edu
am6996@usma.edu

Abstract. The Department of Mathematical Sciences at the United States Military Academy (USMA) is fostering an environment where students and faculty become confident and competent problem solvers. This assessment will reevaluate and update the math core curriculum,s program goals to incorporate the laptop computer, enabling exploration, experimentation, and discovery of mathematical and scientific concepts.

Background and goals

Technology has made a dramatic impact on both education and the role of the educator. Graphing calculators and computer algebra systems have provided the means for students to quickly and easily visualize the mathematics that once took effort, skill, and valuable classroom time. The Calculus Reform movement sought to improve instruction, in part, by taking advantage of these technological resources. Mathematical solutions could now be represented analytically, numerically, and graphically. The shift in pedagogy went from teaching mathematics to teaching mathematical modeling, problem solving, and critical thinking. Ideally the problem solving experiences that students encountered in the classroom were interdisciplinary in nature. Mathematics has truly become the process of transforming a problem into another form in order to gain valuable insight about the original problem.

Portable notebook computers provide an even greater technological resource that has led us to once again reexamine our goals for education. Storage and organization coupled with powerful graphical, analytical, and numerical capabilities allow students to transfer their learning across time and discipline.

The Department of Mathematical Sciences at USMA is committed to providing a dynamic learning environment for both students and faculty to develop self-confidence in their abilities to explore, discover, and apply mathematics in their personal and professional lives. The core math program attempts to expose the importance of mathematics, providing opportunities to solve complex problems. The program is ideally suited and committed to employing emerging technologies to enhance the problem solving process. Since 1986, all students at USMA have been issued desktop computers with a standard suite of software; this year the incoming class of students (class of 2006) will be issued laptop computers with a standard suite of software. The focus of this assessment is to reevaluate the program goals of the math core curriculum and update these goals to incorporate the ability of the laptop computer to not only explore, experiment, and discover mathematical and scientific concepts in the classroom, but also provide a useful medium to build and store a progressive library of their analytical and communicative abilities.

Description

The general educational goal of the United States Military Academy is "to enable its graduates to anticipate and to respond effectively to the uncertainties of a changing technological, social, political, and economic world." The core math program at USMA supports this general educational goal by stressing the need for students to think and act creatively and by developing the skills required to understand and apply mathematical, physical, and computer sciences to reason scientifically, solve quantitative problems, and use technology effectively.

Cadets who successfully complete the core mathematics program should understand the fundamental principles and underlying thought processes of discrete and continuous mathematics, linear and nonlinear mathematics, and deterministic and stochastic mathematics. The core program consists of four semesters of mathematics that every student must study during his/her first two years at USMA. The first course in the core is Discrete Dynamical Systems and an Introduction to Calculus (4.0 credit-hours). The second course is Calculus I and an Introduction to Differential Equations (4.5 CH). The sophomore year's first course is Calculus II (4.5 CH), and the final core course is Probability and Statistics (3 CH). Five learning thread objectives have been established for each core course. They are: Mathematical Modeling, Mathematical Reasoning, Scientific Computing, Communicating Mathematics, and the History of Mathematics. Each core course builds upon these threads in a progressive yet integrated fashion.

The assessment focuses on the following aspects of our core math program:

1. Innovative curriculum, instructional, and assessment strategies brought on by the integration of the laptop computer.
2. Student attainment of departmental goals.

Innovative curriculum and assessment strategies

Projects: In-class problem solving labs serve as a chance for the students to synthesize the material covered in the course over the previous week or two. Students use technology to explore, discover, analyze, and understand the behavior of a mathematical model of a real world phenomenon. Following the classroom experience, students will be given an extension to the problem in which they are required to adapt their model and prepare a written analysis of the extension. Students are given approximately seven to ten days to complete the project. For the most part, these out-of-class projects will be accomplished in groups of two or three. An example of a project is provided in Appendix A. To add realism to the scenario, we create interaction between the model's components by means of extensions that force the students to adapt their model and prepare a written analysis.

Two-day Exams: Assessment of student understanding and problem-solving skills will take place over the course of two days. Paramount in this process is determining what concepts and/or skills we want our students to learn in our core program. We understand that "what you test is what we you get"; therefore, we have adapted our exams to assess these desired concepts and skills. The first day of the exam will be a traditional in-class exam in which students do not have access to technology (calculator or laptop computer). This exam portion focuses on basic fundamental skills and concepts associated with the core mathematics program. Students are also expected to develop mathematical models of real world situations. Upon completion of this portion, students are given a take-home scenario that outlines a real world problem. They have the opportunity to explore the scenario on their own or in groups. Upon arrival in the classroom the next day, the scenario is adapted to allow students to apply their problem-solving skills in a changing environment. An example of a take-home scenario and the adapted scenario is provided in Appendix B of our report on the SAUM website.[1]

Modeling and Inquiry Problems: To continue to develop competent and confident problem solvers, students are not given traditional examinations in the second core mathematics course. Instead, they are assessed with Modeling and Inquiry Problems (MIPs). Each MIP is designed as an in-class "word problem" scenario to engage the student for about 45 minutes in solving an applied problem with differentiable or integral calculus or differential equation methods. The student must effectively communicate the situation, the solution, and then discuss any follow-on scenarios, similar to the Day Two portion outlined above, all in a report format. As an example, a MIP may involve using differential calculus to solve a related rates problem.

The "Situation" portion of the MIP involves transforming the words into a mathematical model that can be solved, by drawing a picture, defining variables with units, determining what information is pertinent, what assumptions should be made, and most importantly, what needs to be found. Finally, the Situation ends with the student stating which method (related rates in this case) will be used to solve the problem. The "Solution" portion involves writing the step-by-step details of the problem and determining what is needed to be found. Any asides or effects of assumptions can be written in as work progresses, and this portion

---

[1] www.maa.org/saum/cases/USMA.html

ends with some numerical value, to include appropriate units. For example, "the rate at which the oil slick approaches the shore is two meters per minute."

The MIP itself has a second paragraph that asks follow-on questions. "Suppose the volume of the oil slick is now doubled. How does that affect your rate?" Or "what is the exact rate the moment the slick reaches the shore?" These follow-on questions prod the student to go back to the method and rework the problem with new information.

The final portion of the MIP write-up is the "Inquiry/Discussion" section. The MIP write-up must be coherent and logical in its flow. Students must tie together the work and stress the solution back in the context of the problem. The Inquiry section is vital in student understanding of the problem. Students do not stop once they determine a numerical answer. They must continue and communicate how that answer relates to the problem, and more importantly, if the answer passes the common sense test.

As of the time of this writing, the third core course has also incorporated MIPs, in addition to traditional exams. The probability and statistics course is considering the use of MIPs in future years. An example of a MIP (focusing on a differential equations problem) is provided in Appendix C of our report on the SAUM website.

Electronic Portfolio: The notebook computer provides a tremendous resource for storage and organization of information. This resource avails the opportunity for students to transfer learning across time and between courses. In the novel, *Harry Potter and the Goblet of Fire*, Dumbledore refers to this capability as a "pensieve."

"At these times," says Dumbledore, indicating the stone basin, "I use the Pensieve. One simply siphons the excess thoughts from one's mind, pours them into a basin, and examines them at one's leisure. It becomes easier to spot patterns and links, you understand, when they are in this form." The portable notebook computer provides the resource for students to create their own pensieve. Creative exercises offer the student exposure to mathematical concepts with the ability to explore their properties, determining patterns and connections which facilitate the process of constructing understanding. Thorough understanding is feasible in either a controlled learning environment or at the student's leisure. Instructors will provide early guidance to incoming students on organizational strategies and file-naming protocol. Informal assessments of a student's electronic portfolio will provide information regarding the ability to understand relationships between mathematical concepts.

Attitude and Perceptions Survey: One tool that will be used to assess if students are confident and competent problem

1. An understanding of mathematics is useful in my everyday life.
2. I believe that mathematics involves exploration and experimentation.
3. I believe that mathematics involves curiosity.
4. I can structure (model) problems mathematically.
5. I am confident in my ability to solve problems using mathematics.
6. Mathematics helps me to think logically.
7. There are many different ways to solve most mathematics problems.
8. I am confident in my ability to communicate mathematics orally.
9. I am confident in my ability to communicate mathematics in writing.
10. I am confident in my ability to transform a word problem into a mathematical expression.
11. I am confident in my ability to transform a mathematical expression into my own words.
12. I believe that mathematics is a language which can be used to describe the world around us.
13. Learning mathematics is an individual responsibility.
14. Mathematics is useful in my other courses.
15. I can use numerical and tabular displays of data to solve problems.
16. I can use graphs and their properties to solve problems.

Figure 1. Questions used in Attitude and Perceptions Survey

solvers in a rapidly changing world is a longitudinal attitude and perceptions survey. Students will be given a series of sixteen common questions upon their arrival at the Academy and as part of a department survey at the conclusion of each of the four core math courses. A comparison of their confidence, attitudes, and perceptions will be made against those students who in prior years took the core math sequence without a laptop computer. The questions used in the survey are provided in Figure 1. Students responded on a Likert-Scale from 1 (strongly disagree) to 5 (strongly agree).

Revisions Based on Initial Experience

The assessment began in the Fall of 2002 and will track students over a period of four semesters. A pilot study was run in the Spring of 2002 and the following lessons were learned.

Student use of computers on exams: In the initial implementation of the two-day exam, students were allowed to use the computer on both days. Many students used their computers as electronic "crib sheets." This problem may be further exacerbated when student computers in the classroom have access to a wireless network. The Day One portion of the exam has been reengineered to assess skills and concepts that do not require technology of any sort.

Electronic imprints of exams: Core math courses are all taught in the first four hours of the day. The students' dorms are all networked and word travels very quickly. It is currently against our policy to prohibit students from talking about exams with students who have not yet taken the exam. Enabling the use of laptops on exams creates a situation in which an imprint of the exam is on some cadet's computer following the first hour of classes. The Day Two portion of the exam, which is designed to test the students' ability to explore mathematics concepts using technology, will be given to all students at the same time, during a common lab period after lunch.

Power: Computer reliability, particularly in the areas of power is an area of concern. Students will be issued a back-up battery for their laptops. It is forecasted that an exchange facility will be available in the academic building for cadets who experience battery problems in the middle of a test.

Findings

Projects: Students overwhelmingly stated that the course projects helped to integrate the material that was taught in the course. The students, ability to incorporate the problem-solving process (i.e., modeling) increased with each successive project.

Two-Day Exams: The two-day exams provided a thorough assessment of the course objectives. Course-end surveys revealed that the students felt that these two-day examinations were fair assessments of the concepts of the course. The technology portion (Day Two) magnified the separation between those who demonstrated proficiency in solving problems using technology and those who didn't; there was no significant in-between group of students.

Electronic Portfolios: Assessment of the electronic portfolios consisted of individual meetings of all students with their individual instructors. The results of these meetings brought out the point that students needed assistance in determining what material should be retained and how it should be kept. Students realized that material in this course would be needed in follow-on courses, so file naming would be key. Guidance was given to students to incorporate a file management system for later use, but no universal scheme was provided; in this manner, students could best determine their own system.

Additional Findings: Unless assessed (tested), the students did not take the opportunity to learn how to effectively use the computer algebra system *Mathematica*. Students embraced the use of the graphing calculator (TI-89) as the preferred problem-solving tool; they overwhelmingly reported that the laptop computer was a hindrance to their learning.

Use of the Findings

Projects: We will continue to use group projects to assess knowledge; however, we will phase the submission of the projects to provide greater feedback and opportunity for growth in problem-solving and communication skills. Our plan is to have students submit the projects as each portion (Introduction, Facts and Assumptions, Analysis, and Recommendations and Conclusions) is completed.

Two-Day Exams: Content on the Day-One (non-technology) portion needs to be more straightforward, emphasizing the concepts we want students to internalize and understand without needing technology. For the Day-Two (technology) portion, questions should be asked to get students to outline and explain their thought processes, identifying possible errant methods. We need to keep in mind that problems with syntax should not lead to severe grade penalties.

Additional Use of the Findings: We are going to introduce graded homework sets designed to demonstrate the advantage of the computer algebra system and the laptop as a problem-solving tool. Use of the graphing calculator will be limited to avoid confusion and overwhelming students with too many technology options. We plan to review course content and remove unessential material, thus providing more lessons for exploration and self-discovery.

Next Steps and Recommendations

The assessment cycle will continue as we implement the changes outlined above into the first course. The majority of students will enter the second core course, Calculus I which will continue the use of laptops. Six Modeling and Inquiry Problems and one project will be used to assess the progress of our students, problem-solving capabilities.

Acknowledgements. We would like to thank the leaders of the Supporting Assessment in Undergraduate Mathematics (SAUM) for their guidance and support. In particular, our team leader, Bernie Madison, has been instrumental in keeping our efforts focused.

References

1. USMA Academic Board and Office of the Dean Staff (1998), *Educating Army Leaders for the 21st Century*, West Point, New York.

2. J. K. Rowling, *Harry Potter and the Goblet of Fire* (Scholastic Trade Publishing, New York, 2000).

Appendix A. A Sample Project

Humanitarian De-mining

Background. The country of Bosnia-Herzegovina has approximately 750,000 land mines that remain in the ground after their war ended in November 1995. The United Nations (UN) has decided to establish a Mine Action Center (MAC) to coordinate efforts to remove the mines. You are serving as a U.S. military liaison to the director of the UN-MAC.

The UN-MAC will initially have 1000 trained humanitarian de-miners working in country. Each of these trained personnel can remove 65 mines per week during normal operations. Unfortunately, there is a rebel force of about 8,000 soldiers that opposes the UN-MAC's efforts to support the legitimate government of Bosnia-Herzegovina. They conduct two major activities to oppose the UN-MAC: killing the de-miners and emplacing more mines. They terrorize the de-miners, killing 1 de-miner for every 1,000 rebels each week. However, due to poor training and funding, each of these soldiers can only emplace an average of 5 additional mines per week.

Meanwhile, the accidental destruction of the mines maim and kill some of both the de-miners and the rebel forces. For every 1,000,000 mines, 1 de-miner is permanently disabled or killed each week. The mines have the exact same quantitative impact on the rebel forces.

Modeling and Analysis. Your current goal is to determine the outcome of the UN-MAC's efforts, given the current resources and operational environment.

1. Model the strength of the de-mining organization, the rebels, and the number of mines in the ground. Ensure you define your variables and domain and state any initial conditions and assumptions.
2. Write the system of equations in matrix form
   $$A(n+1) = R * A(n).$$
3. If the interaction between the rebels and de-miners as well as their respective efforts to affect the minefields remain constant, what happens during the first five years of operations?
4. Graphically display your results. Ensure you display your results for each of the three entities you model.
5. What is the equilibrium vector, D or Ae, for this system? Is it realistic?
6. The General and Particular Solution for the new system of DDS's using eigenvalue and eigenvector decomposition.

Extensions

Better estimate on casualties. Suppose we receive more accurate data on the casualties due to mines; it may (or may not) change part of your model. Better estimates show that for every 100,000 mines, 2 de-miners are permanently disabled or killed each week. The mines have the exact same quantitative impact on the rebel forces.

Other minefield losses. Other factors take their toll on the number of emplaced mines as well. Weather and terrain cause some of the mines to self-destruct, and civilians occasionally detonate mines. Approximately 1% of the mines are lost to these other factors each week.

Natural attrition of forces. Due to other medical problems, infighting, and desertion, the rebel forces lose 4% of their force from one week to the next. The de-miners have a higher attrition due to morale problems; they lose 5% of their personnel from one week to the next.

Recruiting efforts. Both the rebel forces and the de-miners recruit others to help. Each week, the rebels are able to recruit an additional 10 soldiers. Meanwhile, the UN-MAC is less successful. They only manage to recruit an additional 5 de-miners each week.

Project Report

For the project, your report should address the following at a minimum:

1. Executive Summary in memo format that summarizes your research.
2. The purpose of the report.
3. Facts bearing on the problem.
4. Assumptions made in your model, as well as the viability of these assumptions.

5. An analysis detailing:
   a. The equilibrium vector, D or Ae, for the system and discuss its relevance.
   b. The General and Particular Solution for the new system of DDS's using eigenvalue and eigenvector decomposition.
   c. A description of what is happening to each of the entities being modeled during the first five years of operations.
6. The director of the UN-MAC also wants your recommendation on the following:
   a. If the de-mining effort is going to be successful within the first five years, when will it succeed in eradicating all mines? If the de-mining effort is not going to be successful, determine the minimum number of weekly de-mining recruits needed to remove all mines within five years of operations.
   b. Describe at least one other strategy the UN-MAC can employ to improve its efforts to eradicate all of the mines. Quantify this strategy within a mathematical model and show the improvement (graphically, numerically, analytically, etc.).
7. Discussion of the results.
   a. Reflect on your assumptions and discuss what might happen if one or more of the assumptions were not valid.
   b. Integrate graphs and tables into your report, discuss them, and be sure to label them correctly.
8. Conclusion and Recommendations.

A Comprehensive Assessment Program — Three Years Later

Kenneth B. Hannsgen
Department of Mathematics
Virginia Tech
Blacksburg, VA
hannsgen@vt.edu

Abstract. Recent progress is summarized in a departmental program of data collection and analysis at Virginia Tech. The assessment program and this report focus on service/general education courses in mathematics. Sketches of particular studies illustrate the developing awareness of the possibilities and limitations of assessment. Impact on decision-making in this large mathematics department is emphasized.

Background and Goals

Virginia Tech is a land grant, state, Type I research university, with an overall enrollment of over 25,000 students. The semester enrollment in courses offered by the mathematics department averages 8,000–12,000 students. About 90% of these students are registered in four 1000 to 2000 level calculus sequences designed to meet general education requirements as well as the specialized needs of students in i) math, physical sciences and engineering, ii) business, iii) life sciences, and iv) architecture.

Olin and Scruggs [1][1] reported on the comprehensive assessment program that Olin, as department head, had initiated in 1995. They discussed the ways in which a comprehensive assessment program helps the department to meet its responsibilities and document its achievements. They described collection and analysis of data to measure academic success across the service course spectrum, evaluation of several innovative programs, and monitoring of grading equity across course sections. Outcomes through 1997 were reported. We focus on recent developments in several of these areas.

In contrast to a carefully planned and controlled pilot study, this assessment program is an attempt at department-wide accounting. It was instituted by the department head and has been carried out "on the fly" by faculty members who are not experts and who have other major commitments and interests. To the extent that this situation is characteristic of what might take place in a large, research-oriented department, we hope that this report will help planners who embark on similar efforts. We emphasize the impact of the program on the department rather than the detailed content of any particular study.

Description and Findings

Data management. The process of data collection and analysis described by Olin and Scruggs has survived one change of department head and two changes of assessment coordinator. Mr. Kevin Bradley, a graduate student in the Department of Psychology, completed a two-year term in the latter job in 2002. A four-person faculty committee, chaired by the author, oversaw his work. We collected data on courses, students, tests, and grades. One of Bradley's main accomplishments was to organize all of this into a comprehensive, easily accessed 10-year database, to which we add new data each semester.

---

[1] www.maa. org/saum/maanotes49/224.html.

Common final exams. For each large multisection course, a three-person committee, consisting of recent but not current teachers of that course, writes a one-hour test that makes up half of the final exam. Considerations of grading effort and data management restrict us to multiple-choice format for assessment on this scale. Course coordinators evaluate the results and write short reports on student success in relation to our lists of course goals, and on the quality of the exam questions. They suggest improvements for the test and the course. This is the closest we normally come to a complete assessment cycle. The effort requires logistical coordination, as well as the participation of dozens of faculty, many of whom are uninterested in or skeptical about assessment.

A review of the physical science calculus sequence, conducted in 2002, confirmed that the system does not yet provide useful formative assessment. Different committees write the tests each semester, with the focus on writing an accurate final exam, useful for grading purposes. Fitting the test to the needs of long-term assessment is a secondary consideration at best, and exam writers rarely consult the historical record of previous exams and reports. Beyond this, our original goal lists have turned out to be too vague, and the connection between particular questions and particular goals is ambiguous at best. These problems also reflect the difficulty of finding agreement across the faculty on what the learning goals are and on whether particular test results demonstrate that they have been achieved.

We will next attempt, on a pilot basis, to reformulate the learning goals in a way that is more closely tied to specific test questions. (A natural pilot course is first-year integral calculus for the physical sciences, where we have some experience with competency quizzes.) For example, a goal concerning integration might indicate a difficulty level and specify definite integration rather than antiderivatives. A goal on Simpson's rule would specify whether a table of values or a formula is given. Tests will include a core of questions from a standard pool, so that test construction will be more centralized. There is a danger that we will overemphasize mechanical skills, since the variety of possible questions grows rapidly as one goes beyond these skills. Still, there is a consensus in the department that we need some standard barometer of student achievement.

Academic measurement and technology. Teaching innovations involving technology, including online instruction in a few courses, raise serious and controversial questions about learning achievement. The assessment program has made important contributions here.

The first course in the life sciences sequence (1015, precalculus) is given entirely online, but at first we included one weekly live class session to help orient students to the course and discuss applications. Was this live section worth its cost? In Fall 2000, students with relatively strong math skills, as indicated by SAT scores and high school grades, were given the option of taking the course with and without the live session. We compared grades in the course and downstream in the succeeding courses. The placement method had been designed with pedagogical aims rather than as a randomized trial for measurement purposes. Bradley's study used statistical means to compensate in part for these sampling problems. While the live-session students showed a slight advantage in final grades (attributable largely to differences in homework scores), there was no difference between live-session and "independent" students in final exam scores. Moreover, a follow-up study of students who went on to the next course (1016, differential calculus) showed no difference in pass rates between students from the two 1015 groups. Among the few students who crossed over from 1015 to differential calculus in the physical science/engineering sequence, those from the independent 1015 group actually did better. Since the main purpose of 1015 is preparation for subsequent courses, these results gave convincing evidence that the live sessions were not cost effective.

In 1997, the two-credit first-year course on linear algebra[2] shifted to an all-online mode for the entire 2000-student annual enrollment, excluding honors sections, using locally written software. The change yielded dramatic cost reductions, enabling the department to shift resources to other areas. (Reports on implementation, learning outcomes, and cost savings appear on the website of Pew Program in Course Redesign,[3] which supported the development project. A more detailed appendix on assessment is available on request from the author.) Our assessment statistics documented increases in the percentages of students achieving at least a C or at least a D– in the course. General grade levels remained steady through the change. Final exam results indicated that the topical area of eigenvalues and eigenvectors needed more attention, but no clear trend emerged in overall scores. This was not a controlled study, and the assessment results are open to varying interpretations; in particular, the syllabus and tests changed as course developers learned what they could do with the online medium. On balance, however, these results showed that expanded online efforts offer a reasonable way for the department to control costs and maintain effectiveness in lower-level courses, subject to the availability of startup

---

[2] course-delivery.emporium.math.vt.edu/courses/math1114/index.html

[3] www.center.rpi.edu/PewGrant/RD1award/VA.html

funding. In 2002–03 a second course of the life sciences sequence, differential calculus, went completely online in this way. In addition, the sequences for business and physical sciences now include some use of the online utilities.

The linear algebra course has also provided feedback on student attitudes, through an online survey that we conduct near the end of each semester. This is a long survey, with participation rates of over 90%. Questions address students' work habits and their satisfaction with various learning resources. For example, the fraction of students who agreed that the online tutorials "explained concepts well" increased from 71% to 88% over the years. The surveys help the department to track the performance of tutoring staff at the Math Emporium,[4] a large learning center where students do much of their work and testing in the course.

Special calculus sections. As reported in [1], Virginia Tech instituted an "emerging scholars" program (ESP) in 1996, modeled loosely on successful programs at other institutions, such as the University of California Davis.[5] Early success led to an expansion of the program to cover nearly all students predicted to be at risk in first-year engineering calculus. (In contrast to the Davis program and others, then, Virginia Tech Emerging Scholars was not restricted to a small, highly motivated subset of the student population.) As pilot funding from the university administration began to run out, the department turned to assessment results in order to decide whether to press for continuation of the program. Because of year-to-year changes and other factors, the data from this program were very messy, and Bradley's study went through numerous revisions as he and the assessment committee wrestled with questions and assumptions. The final report uses a variety of statistical stratagems to sort out the picture, and there are conflicting outcomes in several cases. Overall, attending the additional ESP hours helped the at-risk students to survive the first-year courses, but there was no carryover benefit (and perhaps even some negative effect) in the second-year courses (differential equations and multivariable calculus). For example, out of $n = 2003$ pairs of ESP and non-ESP students enrolled in the course between 1993 and 2001, matched on the basis of SAT scores and high school grades, 69% of the ESP students versus 60% of the non-ESP students successfully completed the first year of calculus, but only 27% of the ESP students versus 32% of the non-ESP students successfully completed the second year; these trends were confirmed in several other outcome measures. These results

supported a decision not to commit further resources to the program in this form.

Insights

The experience at Virginia Tech is an example of learning the hard way. As traditional mathematics faculty, we are concerned and careful about evaluation of students, but systematic educational assessment is an unfamiliar and in some ways uncomfortable experience. We began with a new department head's need to understand what was happening in our classes (many of which involved experimentation) and to demonstrate the value of what we do to internal and external providers of funding for major changes. We collected data and compiled results, and the process began to yield some insights into the successes and challenges in our program. The availability of an organized body of data has enabled the department to respond to subsequent requests for assessment.

We have found that the numbers do not provide simple, conclusive answers. Results can shed light on outcomes and help in making decisions, but it is rare to find a smoking gun that resolves controversies or overcomes strongly held beliefs about pedagogy. We have gotten some useful large-scale information, but we have as yet found little information that suggests how we need to shift emphasis in the syllabus of any one class. We continue to seek better ways to incorporate assessment into the improvement process.

Acknowledgement. The author thanks Professor Peter Haskell for extensive comments and suggestions.

Reference

1. Olin, R. and Scruggs, L. "A comprehensive, proactive assessment program," in Gold, B., Keith, S. Z., and Marion, W. A. eds. *Assessment Practices in Undergraduate Mathematics,* MAA Notes #49. Washington, DC: Mathematical Association of America, 1999.

---

[4] www.emporium.vt.edu/

[5] www.math.ucdavis.edu/~kouba/KoubaHomepageDIRECTORY/ESP.html

Assessment of a New American Program in the Middle East

Thomas W. Rishel
Weill Cornell Medical College
Doha, Qatar
rishel@math.cornell.edu

Abstract. This paper provides an assessment of a new American-style program in a medical college in the Middle Eastern nation of Qatar. It includes a consideration of the appropriateness of the mathematical curriculum to Middle Eastern students and to a medical program; an evaluation of examinations and projects with respect to critical thinking; and a post-semester student survey to ascertain further needs. Conclusions apply both specifically to this program and generally to how mathematics fits into all medical education.

Background

In August 2002, Cornell University took the daring, unusual, and in some circles controversial step of opening a branch campus of its medical school in the city of Doha in the Middle Eastern nation of Qatar. About 790,000 people live in the country, almost all of them residing in the city of Doha itself. A few other towns in the Connecticut-sized nation support either the fishing or the liquid natural gas industry. Eighty percent of the residents of Qatar are from outside the country ("ex-pats"). Most come from the Indian subcontinent, with other residents from the Philippines, Malaysia, Egypt, Syria and Lebanon, and some from Europe and the Americas.

An Emir, His Highness Sheikh Hamad bin Khalifa al-Thani, governs Qatar. A new constitution has just been approved which supports a parliamentary form of government, and all Qatari citizens will be eligible to run for and to elect the parliament. The Emir's wife, Her Highness Sheikha Mozah, is the director of the Qatar Foundation for Education, Science and Community Development, a private foundation established in 1995 that has invited Cornell to found our medical college in Qatar.

Assessment of Candidates for Admission

As an essential condition of its involvement in the WCMC-Q project, the Cornell administration requires that all students at the Doha campus receive a "full Cornell education." Thus the program is designed as a six-year premedical and medical education, from freshman to medical doctor, with criteria and standards equal to those asked of students in the United States. Students are selected by the Cornell admissions office, not through the Qatar Foundation. SAT and TOEFL scores comprise a large part of the criteria. Further, a candidate's enrollment in the undergraduate program does not guarantee automatic admission to the graduate medical school; every candidate must pass another rigorous second-year review by a separate admissions board, which will assess undergraduate grades and scores on the MCAT.

Most of the students come from outside the United States; hence admissions folders need to be "translated" between nations. For instance, many schools in Lebanon follow a French-style tradition, while those in Kenya have a British curriculum. One effect of these differences is that WCMC-Q has hired an admissions officer with extensive experience in international medical education. Another is that most faculty members are now quite involved in the admissions process. The pre-interview of candidates becomes an important determiner of admission, and we therefore have designed a variety of questions specific to determining whether individual students will be successful in the program.

I met with four candidates the first year; this year, I have seen five more. In each interview, my questions have to do with previous experience, especially in science; about any work experience they may have had, especially in hospitals or clinics; about the types of medicine that interest them; and about how they see themselves attaining their goals over the next few years.

Numbers. In the first year, WCMC-Q offered admission to thirty students. Twenty-six arrived. One student, a Qatari, was offered late admission to a medical school outside the country; he chose to leave WCMC-Q during the third week of class. Since then we have not lost any more students.

The Original Plan

When I was hired to teach calculus at WCMC-Q[1], I was given two not quite overlapping, perhaps contradictory, goals:

- Replicate the traditional Cornell curriculum in mathematics,
- Add to that program according to the medical education and cultural needs of the students.

In considering the above, I decided that the second of the two charges was the more pertinent. My reasoning was as follows:

First, Cornell has no premedical major, as such. Students who wish to go on to medical school major in biology or chemistry—or philosophy, say, or mathematics. Thus, strictly speaking, Cornell has no fixed "calculus for medicine" or "calculus for biology" courses. The course I was asked to teach is called "Math 106: Calculus for the Social and Life Sciences" at the Ithaca campus, renamed Math 104 for Doha.

Second, the text for the course is one that includes some problems from the life sciences, but these problems are usually ones designed to fit the topic being taught. For instance, in a chapter on derivatives of polynomial functions, one of the problems might ask students to graph a cardiac output function:

$$g(x) = -0.006x^4 + 0.140x^3 - 0.53x^2 + 1.79x,$$

then use the derivative to find the maximum cardiac output. While it is worthwhile to get students thinking that deriva-

tives of polynomials might have some use in measuring cardiac output, no significant discussion was spent on why such a function really related to actual cardiac activity. Thus I decided that if I were to really make the course relevant to the future needs of the students, I would have to find and discuss "real world" examples from biology and medicine. Of course, this could be a dangerous course of action for me given what I know about biology and medicine.

Assessment I — Change on the Run

Math 106 is a basic course in calculus that attempts to incorporate a large number of topics into one semester. The course begins with an algebra review, followed by introductory differential and integral calculus, including word problems and computations of areas and volumes. It then has a touch of partial derivatives, and finishes with two weeks worth of differential equations. (My syllabus is posted as Appendix A in the web version of this report.[2]) In the absence of any experience with students at WCMC-Q, or in the Middle East, I decided to use the main campus' Math 106 syllabus as my benchmark, and especially not to skip the algebra and trigonometry review. I soon found, however, that only one of our students could have been termed deficient in those skills. Of course, by the time I had realized this, we had already gone through five class days of discussion of functions, polynomials, logs and exponentials.

When we reached topics in differential calculus, the students again assured me that they had "seen it all." I am used to this reaction, however. In my twenty-seven years at Cornell's main campus, it almost always turned out that each student had seen about 50% of the material. However, each had seen a different 50% at a different emphasis with easier exercises and probably without word problems or related rates problems. This time, my assumption proved correct; this material and subsequent topics proved to be the proper mathematics at the proper pace.

There was another even more important reason for teaching the standard college-level course at the standard rate. As I soon learned, students at WCMC-Q came from a background where quick recall of factual information was central to the educational process; in fact, that often appears to be the entire educational process. Thus it was often necessary for me to emphasize the concepts of calculus, the "Why does this work?" aspects, rather than just teaching "How to solve it" over again. Let me add here that most students at the main campus exhibited the same behavior—it was just more noticeable here in Doha.

---

[1] I use the first person throughout. With the exception of a teaching assistant, I am the entire Department of Mathematics — for better or worse. Faculty meetings are easy.

---

[2] www.maa.org/saum/cases/WeillMed-A.html

As the semester progressed, I did make a number of changes, not in the material itself so much as in the ways in which I approached it. Among these changes:

- Additional examples from biology, chemistry and physics.
- A slightly different style for examinations.
- More quizzes and oral work.
- Some changes in the "Math Review" sessions (as explained below).
- A final project as an alternative to the final exam.

To expand on each of these topics:

My additional examples often came from textbooks like those of Adler [1] and Edelstein-Keshet [4]. A few others were mathematical expansions of discussions from the instructors in the biology, chemistry, and physics courses that I attended. I found it invaluable to see how mathematics is actually used in the biology or chemistry classrooms, and I highly recommend that pre-medical and medical faculty collaborate this way.

I also adapted materials from some "biology and mathematics" web sites. Many of these examples needed to be revised for the audience, however, either because they deleted much of the mathematical aspects of the topic or because they brought in methods that are too advanced for the students' level. I also included a number of biological examples in the differential equations notes that I wrote for the last two weeks of the course. A reference to these notes is in the bibliography [7] and I can send them on request.

My lectures and examinations also changed in that I occasionally offered examples concentrating on the conceptual aspects of topics, and asked questions based on the above examples. For instance, in one examination I proposed a mathematical model for neuron activity and then asked the students to use it to find the maximum ratio of axon to sheath in this model.

A second method I used on examinations was to give multi-part questions whose last part or two asked students to answer questions like "What should this look like?" "Is the model realistic? Why or why not?" "What does the mathematical answer from the previous part of this problem mean to a biologist?" "How do you interpret the graph?"

I gave more quizzes in class because I found out that the typical WCMC-Q students were strongly inclined to operate in strict crisis mode; that is, they would study mathematics only during the week that the exam was coming, then completely ignore my subject until the week of my next exam. (I will say more about this in the "cultural cues" section coming up.) Unfortunately, I didn't rediscover the usefulness of frequent quizzes until late in the semester, but I won't forget for next year's class.

Differential Models of Tumor Growth and Repair
Measurement of Cardiac Output Using the Dye Dilution Method
Calculus and Chemical Kinetics of Reactions
Something About Dialysis
Autoimmune Diseases: Systemic Lupus Erythematosis
Enzymatic Reactions
Exponential and Differential Equations of Tumor Growth and Cure Probabilities
Poiseulle's Law
Oral and Intravenous Drug Intake
The Nitrogen Washout Technique for Pulmonary Function
The Shuttle Problem

Figure 1. Final Projects

There are a number of research papers supporting this strategy. Frequent testing has many important benefits for the learner, in that it encourages regular study habits and decreases cramming. For details, see references [3] and [6]. Test anxiety is reduced, according to [3]. Further, research shows that students favor frequent testing [2], and that these frequent tests also consolidate learning [3]. Finally, a study by Spitzer [9] shows the beneficial effects of giving examinations very soon after instruction.

When I was at the main campus in Ithaca, I found that my Sunday evening "Math Review" study halls were extremely popular with students. (For more detail on these, see Lewin and Rishel [5].) At WCMC-Q, however, with students in "crisis mode" all the time, I found that most were unlikely to use these sessions in the intended manner. Thus, I have decided that in the coming year I will use these periods for algebra and trigonometry review in the first few weeks, followed with some additional material on statistics. Only during examination weeks will I use these sessions as real review for the current calculus materials.

About eight weeks into the semester, I decided to offer the students the option of a final project as an alternative to the final examination. The project was described as a roughly five-page paper on a topic from biology or chemistry using significant mathematics from the calculus course. Twelve students opted for projects; some topics are listed in Figure 1.

Students who chose projects did so for a variety of reasons: "It will be more interesting, challenging, relevant to medicine." "I need to learn how to write papers." "I'm better at writing than math." Those who opted for the exams said: "It will be shorter/take less time to prepare." "I've been doing well at exams; why change now?" "I started to do a project, but it was taking too much time." "I just couldn't get my act together."

From my perspective, the papers took a great deal of time. For instance, when the students found out that I was coming to school at 7 AM, they started to do the same to "talk to me about the project." This was very nice, of course, but it also meant that I could no longer do some of my other work at school, even on weekends.

In investigating final grades, I found that they averaged to the same letter grade regardless of whether students took the examination or did the project, in the following sense:

- The 13 students who took the exam averaged B for a final grade.
- The 12 students who wrote projects averaged B for the final grade.
- While some individual student grades on final projects deviated up or down one letter grade from prelim grades, the same was true for the students who took the final exam.
- The final examination changed seven of the thirteen grades; four went up, three fell.
- The final project changed five grades; four rose, one fell.

The last two bullets merit more reflection. Changes in grade could have reflected grading policy, or they could have been caused by the students' desire level. Perhaps those who chose projects put more effort into their work, either because they were interested in it or because they actually did write better. Or maybe those who chose final exams did so because they thought the test would be easier, and then they didn't work as hard. Which possibility is correct? I think it's a combination.

Assessment II — Factoring in Conceptual Knowledge

Upon discovering the students' relative weakness in conceptual knowledge, I revised the curriculum and examinations to place more emphasis on conceptual knowledge of calculus. A sense of my approach to lectures can be obtained from my differential equations notes [7].

I also compared student grades on "factual" versus "conceptual" questions on my examinations. The student average on conceptual questions was 64%; that on factual information was 65%. Of course, again there were open questions. Who decided what a conceptual question was? Well, I did, based on using Benjamin Bloom's model of cognitive levels. Bloom's model is fairly well known in educational circles, less so in mathematical ones; for some details on this model and how I use it in mathematics, look at chapter 27 of my text [8].

Another question: Who graded these exams, and was the grading uniform?

Answer: My TA and I graded all exams, and the grading was "uniform by question"; i.e., he would grade all responses to a particular question, I would grade all of another.

A third question: Were students primed for specific conceptual questions, or were the problems ones they hadn't seen before?

Answer: These were new problems, although of course students needed to use mathematical methods that had been discussed in lecture beforehand.

A final question: What was the relative amount of conceptual questioning?

Answer: Only about 10% of all the examination material was conceptual in nature. This was partly because I didn't arrive at this methodology until the semester was about six weeks old, after the first examination. I will give more conceptual material next year, but will still include only about 20%.

A Cultural Cue

During the second semester, when I was not teaching, two students provided the physics instructor a strong cue as to one of the differences between Middle Eastern and American education and its impact on student development.

One of the students asked the physics instructor, Marco Ameduri, not to show her the grade on her third exam. "I know I did badly," she said, "I just don't want to see the exam." In discussions on this case, Marco and I agreed that he should tell her that she needed to see where she had gone wrong in order to get it right for the final exam. The next time he saw the student, he told this to her. Her response was not what either of us expected, however. "I know perfectly well what I did incorrectly," she responded. "That isn't the reason I don't want to see the exam. The reason is that I am used to having exams that are based totally on memorization, and whenever I see the problems during the test, I immediately try to remember which homework problem it was. It will not be necessary to see the test."

A second student who started slowly in my course confirmed this phenomenon. She came to me to discuss some causes of her improvement. I paraphrase what she said:

When I used to get an exam problem, I would try to remember all the homework, all the lectures, and all the examples I had ever seen. There was so much to think of and remember that it just got confusing. After some bad exams I realized that any problem I would see on the exam wasn't going to be one I had seen before; it would be new, and I would have to solve it from first principles. That was actually easier. I still fall back, however, into the behavior I've been trained to use; I have to consciously remember not to do that.

The above student examples indicate that, even though there are students outside the Middle East who try what might be called the "mental rolodex" method of solving science and mathematics problems, here in Qatar the quick recall of factual information type of examination is so pervasive that these students must be repeatedly and actively discouraged from using it.

Of course, this may bring with it another problem. We are dealing here at WCMC-Q with a group of students who have been extremely effective in the "rolodex" method; to change now could lead to a certain amount of resentment. How this last is handled will be a further question for all faculty to consider.

I will continue to explore cultural cues to pedagogy as my tenure here at WCMC-Q continues.

Assessment III — The Survey

No assessment paper is complete without a student survey. Mine is found in Appendix C of the web version of this report. [3] Here I will offer an analysis of the results.

I did not give this survey until late in the following semester, in April 2003, to give students time to think about whether the mathematics course had been useful. In fact, this last was my first question.

Findings from the survey:

- The only topics that any of the students mentioned as being difficult were word problems and the final differential equations section. No one advocated dropping either, however, and in fact, they all said these were the most useful of the topics they studied. Two students proposed that I simply spend more time on the differential equations section; this looks like an excellent solution to the problem.
- Some students suggested that I collect homework assignments. I think that instead my approach will be to often ask one of "last night's problems" on quizzes.
- They especially liked the real world applications. No one suggested that I do fewer; many asked for more. Many students mentioned "applicability to biology and medicine" as being important in making the mathematics relevant to their experience. "Having problems that are not in the book and then try[ing] to solve them in recitation and then seeing more applications will...make [the course] more interesting."
- Only two students suggested that I add any new topics to the course: Taylor series and line integrals. One of those two students mentioned that the mathematics I taught fit very well with the physics course, which is "all about

derivatives and integrals. Therefore, mathematics has definitely helped us throughout the physics course."
- The students said that there were enough homework problems, and at the proper level. One student suggested that I should make them do the homework by having them "handed in by the students and checked even if not graded."
- Everyone concluded that the examinations were fair and no substantive suggestions for changes were made.
- Although they found the lectures and differential equations notes useful, they were not as enthusiastic about the recitations or the math reviews. Sherwood's lectures were popular, however. (Sherwood is my teddy bear; he gave two guest lectures: one on mechanics of the heart, one on an ecological problem.)

The question that elicited the longest responses was the one about whether their general reasoning skills improved. Everyone responded with an unqualified yes—but all with somewhat different reasons. "My logic has certainly evolved...since medical and biological examples were given," said one. "I liked the questions where we were interpreting the graphs. I think that improved my 'readings' of graphs," said a second. "Yes, due to the applications we were taught," was a third response.

A final comment about the survey:

An underlying theme of the responses was "we wanted you to work us harder. We wanted more homework problems; we wanted more and harder quizzes; we wanted to have more biological problems to take home and try; some of us even wanted a couple more topics in the lectures." [This is not a direct quote; rather, it's a pastiche I have drawn up from the sense of what I heard from the surveys.] Given that I assigned approximately three times as many homework problems as would be assigned in Ithaca, that I covered much more material in differential equations than the students would ever see there, and that I assigned a very labor-intensive final project, I find that most remarkable.

Uses of the Findings

Many changes have already been described, but I will summarize them here.

Rather than discussing algebra and trigonometry formally in the main lectures, I will give a diagnostic on day one and offer review sessions on the first two Wednesdays. At the end, I will carve out extra days for a more careful discussion of differential equations; but I will not add more topics to this section.

Some statistical topics will be added, including a short discussion of hypothesis testing and regression. The reason is that it could be argued that for a doctor an understanding of statistical methodology is at least as important as a sense of how calculus applies to medicine. (I myself would argue that both are essential.) To find time for statistical topics, I will replace last year's algebra review as well as some of the Math Review time.

Biological and chemical examples will be present from the start. This will be easy enough to do if I begin with logarithms and exponentials. Further, my applications-oriented approach will continue all the way through differential equations.

Conceptual questions will enter in right away. "Why does this work?" "What is wrong with the following model?" I have always used quite a few questions in my exposition; even so, they will be increasing in the future.

All examinations will include some questions emphasizing higher-level critical thinking. Many short quizzes will be added, for reasons outlined in earlier sections of this paper.

Projects will continue to be used, with the goals being to enhance writing and reasoning skills, and to increase student awareness of how mathematics can be applied to science and medicine.

Subsequent Plans

Once I start teaching statistics, I will no doubt need to make further changes.

Also, as projects increase, WCMC-Q will be in need of relevant papers and journals. Students here in Qatar do not have as much access to reading materials as they do in the States, and we will have to find ways to bring these to students. Of course, such sources must be appropriate to students' levels; current research papers are unlikely to be usable. It would be wonderful if there were good access to written materials here in Doha, but unfortunately that is not yet the case.

There is a need to teach the students how to read and evaluate scientific literature. "Study skills" in the North American sense are generally low, and I cannot expect others to provide these skills to my students. To this end, I have gathered some materials on such topics as: how to manage time, read textbooks, take notes, and prepare for examinations. I may have to expand these materials.

Conclusion

Teaching at WCMC-Q has been a never-ending revelation to me. As I have mentioned to many people, Qatar is a place where I never in my life expected to go. Not only is the nation of interest to me, but also the cultural and pedagogical questions that I have encountered are ones that can only keep me intrigued for a long time. I have learned at least as much from the students as they have from me, and I have every hope and expectation that the situation will continue.

References

1. Adler, Frederick. *Modeling the Dynamics of Life: Calculus and Probability for Life Scientists*. Pacific Grove, CA: Brooks/Cole, 1998.

2. Bangert-Drowns, R.L., J. Kulik and C. Kulik. "Effects of Frequent Classroom Testing," *Jr Educ Res*, 85 (2), 89–99.

3. Dempster, F.N. "Using Tests to Promote Learning," *Jr Res Dev in Educ*, 25 (4), 213–217.

4. Edelstein-Keshet, L. *Mathematical Models in Biology*. Boston: McGraw Hill, 1988.

5. Lewin, M. and T. Rishel. "Support Systems in Beginning Calculus," *PRIMUS*, V(3), 275–86.

6. Mawhinney, V., D. Bostow, D. Laws, G. Blumenfeld, B. Hopkins. "A Comparison of Students' Studying Behavior Produced by Daily, Weekly, and Three-Week Testing Schedules," *Jr Appl Behav Anal*, 4, 257–264.

7. Rishel, T. *Differential Equations*. [Draft at rishel@math.cornell.edu].

8. Rishel, T. *Teaching First: A Guide for New Mathematicians*. Washington: Mathematical Association of America, 2000.

9. Spitzer, H. "Studies in Retention," *Jr Ed Psych*, 30 (9), 641–656.

Mathematics Programs to Prepare Future Teachers

Assessment of Mathematics Programs to Prepare Future Teachers

Laurie Hopkins
Provost
Columbia College
Columbia, SC
lhopkins@colacoll.edu

One of the driving forces behind the growing interest in assessment is the increased emphasis on accountability in education as a national issue. Programs offering initial certification of teachers were among the first to face this challenge. In fact, as education remains a popular political topic, the pressure on such programs grows and changes in ways that can make the required assessment seem more of a hurdle than an opportunity.

Although state boards of education maintain primary control over the higher education programs that prepare teachers, national accrediting bodies have an increasing influence over state guidelines. The National Council for Accreditation of Teacher Education (NCATE), one of the most respected and influential accrediting agencies, asserts in its mission statement that "through standards that focus on systematic assessment and performance based learning, NCATE encourages accredited institutions to engage in continuous improvement based on accurate and consistent data" (NCATE 2002).

One of the six standards that determine if an institution is compliant with NCATE professional expectations concerns assessment: "The unit has an assessment system that collects and analyzes data on the applicant qualifications, candidate and graduate performance, and unit operations to evaluate and improve the unit and its programs." In explaining how this standard will be evaluated, NCATE documents make it clear that this standard must be taken very seriously, prescribing the criteria for developing an assessment system, quantifying expectations of data collection, analysis and evaluation, and emphasizing the use of data for program improvement. All states have by now issued an assessment mandate for programs that prepare teachers; many states have taken NCATE as the model on which state accreditation is based.

The federal government has also contributed to the pressure on teacher education programs. In the wake of the passage of the *No Child Left Behind Act of 2001*, the United States Department of Education challenged the traditional methods of preparing teachers. At the first annual Teacher Quality Evaluation Conference in June 2002, Secretary of Education Rod Paige released a report with data showing that state certification systems "allow into the classroom too many teachers who lack solid content knowledge of the subjects they will teach." To raise academic standards, the report calls on states "to require prospective teachers to pass rigorous exams in the subjects they plan to teach" and calls on states and institutions of higher education "to revamp their teacher preparation programs and eliminate many of the rigid certification requirements, such as the massive number of methods courses" (US Dept. of Ed., 2002).

Subsequent reports have suggested that alternate paths to certification may be more effective and have called for the creation of a clearinghouse "to identify research-based best practices in relation to ... teacher training and teaching in subject areas" (US Dept. of Ed., 2003).

Given the reality of these pressures, it is safe to assume that all programs involved in teacher certification have some form of assessment measures in place. In many cases, these assessments have been created to satisfy the external requirement and to prove that the program is adequate. In general, program directors do not expect or desire to collect data that will suggest that they change their practice. While assessment measures created in such a climate are not useless, they do not have the potential for creating meaningful conversations and rich learnings that are possible with different approaches. In contrast, the case studies in this section provide examples of institutions that have responded to the national pressure in ways that provide both useful information about their programs and suggestions for ways in which these programs can improve.

One institution, Monmouth University, is establishing a graduate program targeted for two populations simultaneously—middle school teachers who do not have mathematics certification and adults changing careers to become mathematics teachers. The program is an interesting design as the two populations for which it is aimed are not similar in preparation or experience. At Monmouth, the assessment plan is being developed along with the program. Although the concept of developing a program with its assessment plan is not innovative, it is surprising how seldom concurrent development is seen. Gold's description of the findings from the assessment in the early days of the program and the modifications both to the program and to the assessment plan illustrate the profound impact assessment can have in the creation of a program. By using formative and summative assessment measures, Gold has gathered useful data which has shaped the educational experiences that will be provided to the two disparate target populations in the program.

A second institution, University of Texas at Brownsville and Texas Southmost College, has taken one of the stickier assessment issues and turned it into a truly useful tool. The State Board for Educator Certification (SBEC) in Texas administers a state exam, ExCET, to all candidates for certification. The SBEC uses the first year pass rate and the cumulative pass rate over a two-year period to determine when a program will be rated "accredited," "accredited under review," or "not accredited." For institutions in states other than Texas, a similar ranking is done by the "National Report Card" using the results of the PRAXIS II exams. Of course, this kind of published ranking concerns administra-

tors at institutions and usually results in pressure to improve pass rates on the tests. Security concerns meant that the university was not allowed to receive a detailed analysis of the performance of their students on the ExCET, nor were they allowed to see a copy of the exam. They did, however, have access to the ExCET practice test, a test strongly correlated with the ExCET. So the mathematics department adopted the ExCET practice test as a mathematics benchmark exam, administered it to all majors, and used the resulting data to provide tutorials for students in specific areas and to improve course offerings in the department.

The implications for the use of this kind of assessment tool are tremendous and the process is certainly replicable at campuses in other states with other external exams. The important ingredient is access to a sample test that is highly correlated with the national exam. In fact, it seems a logical next step for some of the external producers of exams to begin to provide the relevant information to institutions from the exams themselves. The authors of the UT Brownsville case study enumerate several advantages of using the sample test:

- the test was free to students;
- the test was graded and analyzed by departmental faculty; and
- the students wanted to do well on the test because of the high stakes of the ExCET exam that would follow.

A direct consequence of implementing this assessment was improvement in the measured pass rates. However, the specific data on the kinds of problems students consistently had on the test and the kinds of changes in coursework for the students that might be implicated is potentially even more powerful. The case study concludes before the effects of curriculum changes can be measured.

These case studies provide a creative alternative to the assessment cycle for teacher education, using data to suggest meaningful improvement. Although most programs expect to collect data that confirms their current practice, even excellent programs have the potential to be better. Designing assessment most likely to suggest areas where improvement is appropriate ensures that programs continue to get better over time. Student learning ultimately is the beneficiary of this approach.

References

Gold, Bonnie, Sandra Z. Keith, and William A. Marion, eds. *Assessment Practices in Undergraduate Mathematics*. MAA Notes., Vol. 49. Washington, DC: Mathematical Association of America, 1999. www.maa.org/saum/maanotes49/index.html.

Subcommittee on Assessment, Committee on the Undergraduate Program in Mathematics. *CUPM Guidelines for Assessment of*

Student Learning. (Reprint of "Assessment of Student Learning for Improving the Undergraduate Major in Mathematics.") Washington, DC: Mathematical Association of America, 1995. www.maa.org/saum/maanotes49/279.html.

National Council for Accreditation of Teacher Education. *Professional Standards for the Accreditation of Schools, Colleges, and Departments of Education.* Washington, DC: National Council for Accreditation of Teacher Education, 2002.

United States Department of Education. "Paige Releases Report to Congress that Calls for Overhaul of State Teacher Certification Systems." Washington, DC: 11 June 2002. www.ed.gov/news/pressreleases/2002/06/06112002.html.

United States Department of Education. "U.S. Education Secretary Paige Highlights Department's Highly Qualified Teacher Initiatives." Washington, DC: 3 Sept. 2003. www.ed.gov/news/pressreleases/2003/09/09032003.html.

Assessment in a Middle School Mathematics Teacher Preparation Program

Bonnie Gold
Mathematics Department
Monmouth University
West Long Branch, NJ
bgold@monmouth.edu

Abstract. At a small private comprehensive university in New Jersey, the mathematics department is developing a program for middle-school teachers who do not have mathematics certification, and for adults changing careers to become mathematics teachers. The purpose is to give these teachers a deeper understanding of the mathematics they will teach. The assessment plan is being developed along with the program itself, in line with NCATE standards as well as New Jersey requirements for mathematics certification.

Background and goals

Middle school is often the weakest link in a student's mathematical education. When (as currently) there is a shortage of teachers with certification in mathematics, those with certification usually end up in the high schools. Middle schools often resort to moving an elementary school teacher up to middle school, or have a teacher certified in another subject begin teaching mathematics, often with no more than one college mathematics course as background. Monmouth has been working, for the last several years, on developing a program for these middle-school mathematics teachers. The audience we originally intended the program for was certified teachers who do not have certification in mathematics but who find themselves, at some stage in their career, teaching mathematics at the middle school level. However, our program can also accommodate people entering teaching as a second career (alternate route), and with small modifications, standard undergraduates specifically interested in middle school teaching.

The aim of this program, which will finally consist of six mathematics courses (Foundations of Number Systems, Geometry, Discrete Mathematics and Problem Solving, Probability and Statistics, Foundations of Algebra, and History of Mathematics) is to give teachers a deeper understanding of the middle-school mathematics through connections with mathematics normally taught at the undergraduate level. (The latter is partly because the state requires courses that count toward certification to be courses that are part of a major in the subject.) We chose the courses because together they cover, at a deeper level, the content of middle school mathematics courses. The Number Systems course gives teachers a deeper understanding of the natural numbers, integers, rational numbers, and to a lesser extent the real and complex numbers, which they need to teach middle school students effectively about decimals, percents, proportion, etc. Many middle school students take prealgebra in middle school, and some take first year algebra as well; the Foundations of Algebra course (which looks primarily at polynomial rings as the context in which middle school algebra takes place) makes the teachers better prepared to teach these courses. Current middle school texts include a substantial amount of geometry, a bit of probability, a fair amount of descriptive statistics, and a lot of problem-solving activities. The history of mathematics course gives teachers a sense of how all of this was developed and how the different subjects interrelate.

We try to keep the courses as independent of each other as possible, to allow teachers to enter the program any semester. On the other hand, we need to ensure that, by completing the six-course sequence, the teachers will be well-prepared, both in content and skills, to teach middle school mathematics. Since we accept students in the courses as long as they completed whatever mathematics they were required to for their undergraduate degrees (which may be as little as one college algebra course), for some of the students in the program this sequence is quite a stretch. However, at least for the students we have had so far, many are quite mathematically talented, and all are willing to work considerably harder than many of our undergraduates. As of Spring 2004 we have only taught the Number Systems and Discrete Mathematics courses (both of which I developed and taught), but the Geometry and Probability and Statistics courses, which my colleagues Lynn Bodner and David Sze (respectively) are developing, will be offered during the 2004–2005 academic year. The final two courses will be offered the following year.

These courses are strictly content courses — students get their methods courses from our Education School — but we are trying, in the courses, to *model* both appropriate teaching methods and a range of assessment methods, both formative and summative. Although this program is aimed at students who already have bachelors degrees, it could easily be adapted to form a concentration at the undergraduate level.

Figure 1 provides a detailed statement of the general program goals. These are then detailed further in each individual course's Course Objectives and Expected Learner Outcomes. (The Course Objectives say what experiences the teacher will provide, the Expected Learner Outcomes, what the student should be able to do as a result.) One example of these (for Discrete Mathematics and Problem Solving) is given in detail in Appendix A.

Developing the assessment program

Because I am leading the development of this program and have considerable experience with assessment, we have been developing the assessment plan along with the program. To prepare for developing the program, I participated in a PMET (Preparing Mathematicians to Educate Teachers) workshop in June, 2002 and 2003. As we develop the courses, we examine carefully several series of middle school texts developed in the last ten years with National Science Foundation funding, to see what topics are covered. (These are *Connected Mathematics*, Prentice Hall; *Mathematics in Context*, Holt, Reinhart & Winston; *MathScape*, Glencoe/McGraw-Hill; and *MATHThematics*, McDougal Littell; information about all four can be found at the Show-Me Center's website.[1])

One difficulty in developing this program is the lack of appropriate textbooks for this audience. Most texts directed

- Help middle school teachers develop a deeper understanding of the mathematics they teach.
- Give middle school teachers an understanding of the relationship between the mathematics taught in middle school and undergraduate mathematics.
- Give middle school teachers experience with a range of pedagogical styles and assessment methods.
- Allow middle school teachers to reflect on their own experience as learners of mathematics.
- Help middle school teachers learn to develop lessons which increase their students' critical thinking skills.
- Introduce middle school teachers to appropriate technology available for middle school mathematics students.

Figure 1. Program Goals

at elementary school teachers are less mathematically sophisticated than we want for our students and cover topics in less depth. Texts directed at mathematics majors going through the usual four-year program assume background our students do not yet have.

Our university is planning to apply for NCATE (National Council for Accreditation of Teacher Education) accreditation within a year or two. Thus, our assessment plan must enable us to satisfy NCATE's requirements. NCATE is in the process of adopting new standards, which can be found in the document, *NCATE/NCTM Program Standards (2003)* under Mathematics Education[2]. A description of how these are organized, and of the parts of the standards relevant to our Discrete Mathematics and Problem Solving course, can be found in Appendix B. To develop the program, we have to determine, for each indicator, in which course (or, for the process standards, which courses) students will gain the knowledge required, and how they will demonstrate that they have this knowledge.

Details of the assessment program

Since we are still in the process of developing the program, the primary assessment so far has been in individual courses. We have, of course, given weekly homework assignments and both hour-long tests and final examinations. However, the wide range of both mathematical experience and current mathematical ability in the class, and our desire that all the students (assuming they are doing the work) benefit from the assessment activities, led us to keep the emphasis on these traditional methods relatively small. We use the examinations to test that students have learned the essential mathematical skills and content that we feel every student completing the course must have. For the discrete

---

[1] www.showmecenter.missouri.edu/

[2] www.ncate.org/standard/programstds.htm

mathematics course, this includes being able to use truth tables to determine whether statements are tautologies, use Venn diagrams correctly, formalize an argument in symbolic logic, correctly state definitions and a few theorems, do a proof by induction, prove simple theorems about sets, solve fairly straightforward counting problems, determine whether a graph has assorted properties we had discussed (Euler cycles, planarity, etc.), and use Euler's formula.

However, because our goals are considerably broader than simply teaching certain mathematical topics—in particular, in each course we want to make progress on the NCATE process standards as well—we use quite a range of other assessment tools.

Summative assessment activities:

Portfolio of activities for a middle school class. We start each class (after we have gone over homework from the previous class; class meets once a week for 2 1/2 hours) with a problem taken from a middle-school text to introduce the topic of the day. We then cover the day's topic by a combination of interactive lecture and student activities. In principle, students spend the last fifteen minutes of class working in pairs, developing an activity for a class (middle school unless they are definitely planning to teach at the high school level), based on the topic of the lesson. (In practice, in many class meetings time runs out and this activity becomes an additional homework assignment.) The activity may be something quite brief, which might take their students only five or ten minutes, and does not have to be in finished form. It may be a description of what they would do with the class, or it may be a handout or worksheet, etc. We collect and comment on these portfolios at midterm and at the end of the semester. They are assessed holistically (and commented on extensively), based on how appropriate they are for the grade level they'd be used with, how well and correctly the mathematical language and concepts are used, and how well the activities would help students develop critical thinking skills.

Computer labs. In each course so far, we have done two computer labs. Our criteria for good software are that it be free or nearly so, that it work with current computers, and that it have activities that both the teachers and their students could benefit from. (One exception may be Geometer's Sketchpad, which is reasonable for schools to buy.) The labs for our first two courses can be found on my web pages for the courses. (See links to MA 500-level courses on my web page;[3] the labs are linked to their day on the course syllabus.)

The first lab in the Discrete Mathematics course uses several pieces of software developed at the University of Arizona quite a few years ago, prior to Windows and even prior to computer mice being common. So the programs are rather awkward to a modern user.[4] They are free, of course, and they are excellent in their conception and keep our students engaged. I have not tried them with middle school students, but I think they would work well at that level as well. We also used other software found in the software section of the Math Archives under Discrete Mathematics.[5] For the Number Systems course, we used a number of NCTM's e-lluminations,[6] as well as Excel and Maple.

Curriculum project. A few weeks prior to the end of the semester, students hand in a curriculum project. This must be a 3–5 day unit for a middle- or high-school class, based on some of the ideas we have studied during the semester. It must include detailed learning goals for the activity, an overview of what they would do with the class, detailed lesson plans for each day, worksheets, handouts, overheads and/or computer labs, an assessment plan for the activity, and a description of how the activity is related to what we have done in class. They are allowed to include some of the activities from their portfolios as part of their projects. The grading rubric is handed out with the project assignment.

Formative assessment activities

Students are expected to write each week in reflective journals about their struggles with learning mathematics. Often as much as half of each class is spent on discussion of their difficulties with the concepts, how it relates to what they will be teaching, etc. We spend some time in class discussing school-level problems and the mathematical difficulties involved in teaching the material. For example, in the Number Systems course, when we were discussing the rational numbers, I took eight word problems from the text we use in our undergraduate course for future elementary school teachers, had the students make up a simpler problem and a more difficult problem of the same sort, and asked them to decide what properties of the rational numbers a student would need to understand in order to solve the problem.

Findings and success factors

The students in the two courses I have taught so far have probably been much stronger than those we will generally

---

[3] bluehawk.monmouth.edu/~bgold/

[4] www.math.arizona.edu/software/azmath.html

[5] archives.math.utk.edu/

[6] illuminations.nctm.org/index.asp

have once the program is actually attracting in-service middle-school teachers. In the Number Systems course, of the six students, two were sufficiently mathematically talented that I would be happy to recommend them for graduate work in mathematics, and two others were reasonably talented and worked harder than any undergraduate student I have ever had. If they could not solve a problem in a few hours, they would come back to it over and over again until they solved it! The informal peer pressure from these four—their level of discourse in class, the activities they shared—influenced the remaining two to work well above the level they would have in a less enthusiastic class. All felt that there was too much work, but they made enormous progress. Their reflective journals were astonishing for their depth of insight into their learning processes and the detail of what they tried. For example, from one: "The Euclidean algorithm is still fascinating to me as a way to determine the greatest common divisor. At first I did not realize that there could be more than one value of x and y for the Theorem: $by - ax = \gcd(a, b)$. (I had it confused with the fact that $b = aq + r$ is unique. It wasn't until I discussed it with Sandy [another student in the class], that I realized that this must not be the case.) After discussing with Dr. Gold, I started to look at Lemma 1.1.10 and reached understanding by working it through. What Dr. Gold told me worked for my homework problems, but I wanted to understand why we must choose k such that ka will be $> y$ and kb will be $> x$." (We were working with the natural numbers at that point. She then goes on to explain what she found.) In this course, students asked that I make the course available on our course management software (WebCT), because they wanted to be able to interact with the other members of class outside of class hours, and they had found this system efficient for this purpose in their education courses.

Some of my experiences to date may not be typical of what we will find when we teach the course to actual inservice teachers. The quality of the portfolio activities and the course projects varied considerably. In each class I had one student who was currently a teacher in a private school; most of the best activities came from these two students. However, most of the students in the classes were fairly early in their work on their MAT and had little teaching experience. I was surprised to find that several of the students who were themselves most creative mathematically tended to make up rather routine worksheets for use with classes, or to be overly rigid in their lesson plans. This suggests that these activities are very appropriate for the students we intended the courses for, but less appropriate for the actual audience we currently have.

Both classes enjoyed the computer labs, and the activities I had found on the internet stimulated them to look for more. The Number Systems class seemed to find the Maple project less worthwhile. The logic games from the University of Arizona website were particularly intriguing for our students, as were the NCTM activities which were more game-like in nature (the Product Game and Paper Pool). In the Discrete Mathematics course, I should have broken the second lab into two parts, since it was too long, but I had not expected to find so many appropriate activities for this material.

The Discrete Mathematics and Problem Solving course had no students with the mathematical sparkle of the Number Theory class—it was taken by students with a much weaker background—but again there were several students in the class who worked extremely hard and the class as a whole made excellent progress. In both courses, students learned significantly more than most of our undergraduates do in the corresponding courses. I am not sure whether this was due primarily to their maturity, or to their commitment to their future careers, but the classes were a joy to teach. Possibly my choice, in both cases, of textbooks that were a bit too hard for the class (because they were the only texts I could find that covered most of the subjects I wanted to include) was also a factor in the level of effort they put forth. Examinations were the assessment item they were most anxious about, but all showed, on examinations, a good understanding of correct mathematical definitions and an ability to solve problems of the kind we had practiced, although most had significant trouble developing proofs in a timed environment.

If the classes were substantially larger (say, over 15 students), it would be important to develop rubrics to enable more rapid evaluation of the portfolio items and journals. We would not be able to read every contribution of each student, nor would we grade every homework problem submitted. The curriculum projects could be done in groups of two or three. However, the range of assessment items shouldn't prove overwhelming in classes of up to twenty-five or thirty students, as only the homework is collected weekly.

Use of the findings

The amount of time it has taken the very strong students in the Number Systems course to do the problems has made me realize that the course, as it stands, is too ambitious. The topics are about right, but the level of detail is too high for students who have little mathematical background. We will need to find an easier text or write our own materials. Fortunately, there are a few programs that are developing, with NSF support, materials for some of these courses. They were not yet ready when we first offered the courses, but first drafts are now available. The students' relative lack of enthusiasm for the Maple computer lab I did in the Number Systems course, together with the likely unavailability of such software at the middle school level, led me to develop labs in the Discrete Mathematics course that use only software freely available on the web. We will replace Maple with graphing calculators in the future.

We also need to revise the courses, and somewhat the assessment plan, to reflect the likelihood that, at least unless we get funding, most of the students will not already be practicing teachers. We will therefore decrease the emphasis on projects for their students, as they have no active experience with what middle school students can be expected to do. On the other hand, we need to increase the attention paid to pedagogical issues such as typical student errors and confusions, how to recognize them, and how to respond to them.

Next steps and recommendations

One part of the assessment plan still needs to be developed, namely, to examine how the students in the program change as teachers as a result of this program. So far, in each of the courses we have taught, only one of the six students in the class has been an actual classroom teacher, although some others have been working part-time as substitutes. Once the program is in full operation, however, we hope that at least half the students will be current teachers. For these students, we will collect information on their classes at the beginning of the program and again once they finish the program. The materials we will gather will be two assignments or projects they feel proud of using with their students, as well as a videotape of their class. The assignments will be examined for evidence of appropriateness of the problems for the level of student, development of students' critical thinking skills, and use of mathematical understanding in the choice of problems. The videotapes will be examined to see how their increased understanding of the mathematics improves their ability to respond to their students questions and ideas, and whether they use more interactive teaching styles, encourage students to be active learners, etc.

Appendix A. Discrete Mathematics Courses Objectives and Learner Outcomes[7]

(Note: This course will be designated MA 520 once the program is approved; currently we offer it as MA 598, our graduate Special Topics number.)

Course Objectives: This course will

- Give students a deeper understanding of topics in discrete mathematics taught at the K–12 level;
- Introduce students to heuristics for problem-solving;
- Give examples of how to use these concepts to develop classroom materials to enhance student learning;
- Introduce students to mathematical software available at the school level for investigating discrete mathematics;
- Involve students in problem solving activities via exploration and experimentation to allow students to construct (and reconstruct) mathematics understanding and knowledge;
- Encourage visual reasoning as well as symbolic deductive modes of thought (by incorporating models, concrete materials, diagrams and sketches);
- Introduce multiple strategies of approaching problems by discussing and listening to how others think about a concept, problem, or idea;
- Involve students in small group work and cooperative learning;
- Help students become aware of their own mathematical thought processes (and feelings about mathematics) and those of others;
- Introduce students to multiple methods of assessment in mathematics.

Expected Learner Outcomes: Students will develop the ability to:

- Approach problems from multiple perspectives and help their students become better problem solvers;
- Use mathematical language to correctly state mathematical definitions and theorems;
- Begin developing mathematical proofs;
- Use assorted counting techniques to solve problems;
- Work with the assorted concepts from graph theory and apply them to a range of problems, including map coloring, networks, traversing routes.

---

[7] mathserv.monmouth.edu/coursenotes/gold/MA520hom.htm

Appendix B. An Introduction to NCATE Accreditation Requirements

NCATE (National Council for Accreditation of Teacher Education) is in the process of adopting new standards, which can be found in the document, *NCATE/NCTM Program Standards (2003)* under Mathematics Education.[8] These standards are very detailed, and describe both content knowledge and skills that teacher candidates must demonstrate. Some of the skills will be learned in their education courses, but "process standards" 1–6 (knowledge of "mathematical problem solving," "reasoning and proof," "mathematical communication," "mathematical connections," "mathematical representation," and "technology") as well as the "content standards" (there are also "pedagogy" and "field-based experiences" standards) are primarily learned in the mathematics courses. For each standard, several indicators of what it would mean to meet that standard are listed. The process standard indicators are the same for all levels (Elementary Mathematics Specialists, Middle Grades, and Secondary Level), but the content standard indicators become more elaborate the higher the grade level.

As an example of Process Standard indicators, for "Knowledge of Mathematical Problem Solving" the indicators are

1.1 Apply and adapt a variety of appropriate strategies to solve problems.

1.2 Solve problems that arise in mathematics and those involving mathematics in other contexts.

1.3 Build new mathematical knowledge through problem solving.

1.4 Monitor and reflect on the process of mathematical problem solving.

For the middle level, the Content Standard "Knowledge of Discrete Mathematics: Candidates apply the fundamental ideas of discrete mathematics in the formulation and solution of problems," has as indicators

13.1 Demonstrate a conceptual understanding of the fundamental ideas of discrete mathematics such as finite graphs, trees and combinatorics.

13.2 Use technological tools to apply the fundamental concepts of discrete mathematics.

13.3 Demonstrate knowledge of the historical development of discrete mathematics including contributions from diverse cultures.

NCATE's matrix for each of these indicators asks "How do our candidates acquire and demonstrate the knowledge addressed to this standard?" "What evidence supports candidates' knowledge acquisition and performance?" and "What are our findings?"

---

[8] www.ncate.org/standard/programstds.htm

Using Practice Tests in Assessment of Teacher Preparation Programs

Jerzy Mogilski, Jorge E. Navarro,
Zhong L. Xu
Department of Mathematics
University of Texas at Brownsville and
Texas Southmost College
Brownsville, TX
jkm@utb.edu
jnavarro@utb.edu

Abstract. In this article, we show that a properly constructed comprehensive exam may serve as a useful assessment tool of student learning. We discuss the assessment of the teacher preparation program based currently on the detailed item analysis of a multiple-choice exam. The exam was taken by seniors (in exceptional cases by juniors) and graduates during the preparation for the state exam for teacher certification.

Background and goals

In the state of Texas, The State Board for Educator Certification (SBEC) adopts the accreditation standards for programs that prepare educators, and administers the Examination for the Certification of Educators in Texas (ExCET). Program accreditation is based on the ExCET pass rates. SBEC uses two types of rates: *first year pass rate,* which is the pass rates for the test takers taking the test for their first time who were students at that school and *cumulative pass rate,* which is based on the performance over the two-year period. To be rated "Accredited" a program must achieve 70% first-year pass rate or an 80% cumulative pass rate. Otherwise, SBEC rates programs as "Accredited under Review" or "Not Accredited."

The University of Texas at Brownsville is a young university, established in 1992. The current enrollment is 11,000 students. The Department of Mathematics consists of 18 regular faculty members and three lecturers. The department offers the BS degree in mathematics with three tracks of study: a non-teaching degree with a minor, teacher certification grades 8–12, and teacher certification grades 4–8. The majority of mathematics majors choose a teaching career after graduation. A substantial number of them choose the traditional way through the teaching certification program. The Alternative Certification Program and the Deficiency Program provide two non-traditional ways to become a certified teacher with the non-teaching BS degree in mathematics. For this reason most of the mathematics majors take the examination for the Certification of Educators in Texas (ExCET).

Our study, which started in the fall of 1997, was motivated by a major concern of the administration of the university and the department of mathematics about the poor performance of our students in the ExCET for secondary mathematics teachers. According to SBEC the passing rate of UTB students in the mathematics test was below 50%. This five-hour exam consists of 90 to 100 multiple-choice questions covering 41 competencies, which are grouped into five domains: (1) Mathematical Foundations, (2) Algebra, (3) Geometry, (4) Trigonometry, Analytic Geometry, Elementary Analysis, and Calculus, (5) Probability, Statistics, and Discrete Mathematics. In order to understand the reasons for the poor ExCET performance of our students, we analyzed the correlation between the curriculum for the mathematics major and the ExCET competencies. Using this correlation, we conducted a systematic assessment of student learning in mathematics for the teacher preparation program.

133

Details of the Assessment Program

Although the ExCET is very good in evaluating students' knowledge in the field of mathematics, we could not use it in our assessment because the SBEC provides only very limited information about the student's performance. Fortunately, we found another test which served as our measuring tool, the ExCET Practice Test produced for SBEC by committees of Texas educators and National Evaluation System, Inc. This test was strongly correlated with the ExCET and was designed to assist staff at educator preparation programs in providing feedback on candidate performance in relation to the ExCET test framework. According to the State Board for Educator Certification, ability to answer 70% of the questions on the ExCET Practice Test correctly indicates sufficient preparation for the ExCET. Since the fall of 1997, the ExCET Practice Test has been adopted as the Mathematics Benchmark Test, and it has been administered to more than 255 students. We summarize here the results of our five year study.

The Mathematics Benchmark Test consists of 150 questions distributed fairly evenly among the 41 competencies in the ExCET and spread among the five mathematical domains listed above. In addition, the authors of the report have divided the 41 competencies into 20 competency groups to obtain a clear correlation between the competencies in the test and mathematical topics taught in our Math Major program. The questions in the test are in multiple-choice format and typically it takes six hours for a student to complete the test. The test indicates which competency is tested by which questions and, for this reason, serves as an excellent diagnostic tool.

There were two big advantages of using the Benchmark Test. First, it was free to the student. Second, we graded it in our department, and we could perform the item analysis according to our needs. The latter let us pinpoint the areas where the particular student was weak. This information resulted in the short term in tutoring sessions, as well as special study materials we provided to the student, and in the long term, in revising the curriculum and improving teaching. There was also a third, rather significant, advantage: the test gave students an idea of what they will find on the real exam itself, which made it much easier to study for. Since they could not become (or continue as) teachers until they pass this exam, they were highly motivated to do well on it. We used the tutorial sessions to discuss the test with the students and learn more about their perceptions of the problems. In fact, these discussions make our assessment process multidimensional, and they were a very valuable source of information about the typical difficulties that our students had taking the test.

Statistical Findings

Our analysis of the results of the test started from each specific question in the test. Then, we generalized the findings to competencies, and to the groups of competencies. We finalized the analysis by correlating the results with the instructional goals and the courses offered by the department.

Benchmark Results Overall. The data of this study included scores of tests of 255 students with the item analysis by each of the 150 questions. The average percentage of correct answers was 59% and it stayed constant for the period of five years.

Benchmark Results by the Competencies. The 100 questions were grouped in 41 competencies. Of 41 competencies included in the Mathematics Benchmark Test, students scored below 40% on three, between 40% and 49% on another three, between 50% and 59% on fourteen competencies, between 60% and 69% on fifteen, and between 70% and 79% on the remaining six. Unfortunately, on only 51% of the competencies did students score 60% or above. In order to make our data more transparent, we grouped the 41 competencies into 20 competency groups and tabulated the percentage correct for each group for each of the five years of the study. Details are available in Appendices A and B of our complete report on the SAUM website.[1]

Assessment of Mathematics Courses. We developed a correlation table between the 41 benchmark competencies and the mathematics courses. This correlation enabled us to use the benchmark statistical data to evaluate 19 mathematics courses in the program offered by our department. The only courses in which student scores are below 50% on the material covered in these courses are Calculus I and Calculus III. These results are consistent with the five-year average and the findings of the previous assessments. The courses in which student scores are above 50% but below 60%, are Calculus II, Mathematical Statistics, Trigonometry and Discrete Structures. The only course in which students scored above 70% on its contents is Linear Algebra. In the period of five years most of the courses improved except for the Calculus I, II and III. Details of this analysis are available in Appendix C of our website report.

Calculus Questions in Benchmark. Our students performed rather poorly in the calculus part of the ExCET and calculus has been constantly the weakest area in the Benchmark Test during the last five years. The test has 10 questions about

---

[1] www.maa.org/saum/cases/UT-Brownsville.html

basic calculus concepts and topics. The percent of correct answers on this group of questions averaged 41%, ranging from 20% on two, to 57%. The questions which students answered correctly only 20% of the time seem to be basic for any calculus course. The first asks which of the derivatives is used in order to find the inflection points. The second asks for reconstructing distance from the given velocity function. The five-year averages show that the above problem is persistent.

Qualitative Findings

We had a unique opportunity to discuss the questions from the test individually with each student during the "after test" tutorial sessions. These discussions were a very valuable source of information because the students openly described typical difficulties they had taking the test. Our findings indicate that the most noticeable difficulties our students had are:

- seeing information obviously shown in the given stimuli.
- using properly the mathematical formulas which were provided to them.
- seeing connections between the concepts taught in the different mathematical courses.
- interpreting parametric variation in order to see dynamics of functions.
- solving problems out of context even if the problems were easy to solve.
- solving problems by the methods of considering all possibilities, converting and transforming, pictorial representation.
- thinking analytically and independently.
- describing real world relationships using mathematical concepts.
- reading mathematical text.

Use of Findings

Based on our studies, we found effective ways to improve significantly the performance of the students taking the ExCET through workshops, tutoring sessions, study materials and practice tests. We also started to teach new courses specially designed for the teacher preparation program such as "Problem solving and mathematical modeling" and the two-semester course "Survey of mathematical principles and concepts." Based on our assessment the department has recently made several changes in the mathematical curriculum including increasing the number of credit hours of the calculus courses and making Discrete Structures a required course for mathematics majors. We communicated the

Academic Year	97/98	98/99	99/00	00/01	01/02
Number of First Year Takers	14	20	16	20	23
Number of Passed	3	19	14	17	18
First Year Pass Rate	21%	95%	88%	85%	82%

Figure 1. ExCET Passing Rates for the First Year Takers

assessment findings to the faculty of the department of mathematics and we started to work on establishing a new "classroom culture" for learning mathematics which would focus on understanding concepts and developing good problem solving skills.

All these efforts brought very positive results. We brought the ExCET passing rate above 80% in one year and it has remained there since then. In recent years, our students have been very successful in passing the ExCET. The table below shows the passing rates in the ExCET for the last five years.

Although the figures in the above table differ considerably from those in the tables for the practice test there was an improvement in the practice tests over this period as well. This improvement was not as dramatic as in the ExCET. It shows that the results of curriculum changes take more time and that the workshop and practice test activities are quite effective.

Next Steps and Recommendations

It is too early to see the effect of the curriculum changes that we have made. Our department is working on developing a comprehensive plan of assessment of our major. When the assessment is implemented, we hope that it will provide some information about the effectiveness of the new courses.

On the other hand, we can see whether our graduates are successful teachers. It is easy for us to maintain contact with graduates because many of them become mathematics teachers in the local school districts. We see them regularly at the collaborative meetings and professional development activities offered at UTB. Since the school districts encourage the teachers to go into graduate school, many of our graduates take the graduate courses in mathematics as a part of the requirements for MEd in Mathematics Education offered at UTB. Presently, we are in the process of establishing an assessment program based on the professional development activities for teachers and the graduate courses.

Acknowledgments

This study was partially supported by the UTB Research Enhancement Grant and the NASA Minority University Mathematics, Science & Technology Awards for Teacher Education Program.

Undergraduate Major
in Mathematics

Assessing the Undergraduate Major in Mathematics

William A. Marion
Valparaiso University
Valparaiso, IN
bill.marion@valpo.edu

When most faculty hear the word *assessment*, they think about assessment of the major rather than of general education or quantitative literacy. That's exactly what the members of MAA's Committee on the Undergraduate Program in Mathematics (CUPM) thought when a subcommittee on assessment was formed in the summer of 1990. Out of that effort came the report *Assessment of Student Learning for Improving the Undergraduate Major in Mathematics* (CUPM, 1995). That document contained a five-step assessment cycle which could serve as a template for mathematics departments as they struggled to build their own assessment plans.

Subsequently, a number of demonstration projects (case studies) by colleges and universities who "got in the game early" were highlighted in *Assessment Practices in Undergraduate Mathematics* (Gold et al., 1999). Over the past three years additional demonstration projects have been developed as part of the MAA's NSF-funded program "Supporting Assessment in Undergraduate Mathematics" (SAUM). Together, both sets of case studies help point the way to effective means of assessing majors in undergraduate mathematical sciences programs.

An effective assessment plan must be anchored in the department's mission statement. So the natural first step is for faculty to review and update (or if necessary, write) their mission statement. Once that's been accomplished, goals for student learning outcomes can be articulated. Goals are broadly-based descriptions of what competencies or skills students should have after completing the major.

One approach is to ask each faculty member to complete the statement "Upon completion of the major a student will be able to…." The many faculty suggestions should then be pared down to a relatively small number, say from four to six. Several case studies illustrate this process, including the Colorado School of Mines (p. 149), Columbia College (Hopkins 1999), Keene State College (p. 157), St. Mary's College of Indiana (Peltier 1999), Saint Peter's College (p. 183), and South Dakota State University (p. 191).

Once agreement is reached on goals, it will be easier to develop student learning objectives. The objectives themselves should not be stated broadly, but should be thought of in terms of measurable outcomes. That is, one needs to think about how one knows when a goal has been achieved (or not). Typically, a number of learning objectives must be developed for each goal. Case studies that illustrate this process include Colorado School of Mines (p. 149), Columbia College (Hopkins 1999), Mary Washington College (Sheckels 1999), North Dakota State University (p. 93), and University of Arkansas, Little Rock (p. 201).

When writing a learning objective, one should have some notion of how to measure it. One advantageous approach is to list a variety of possible assessment tools and determine which instruments would be appropriate for measuring which objectives. This does not mean that the department will have settled on which measures they are going to use, only that serious consideration will have been given to articulating objectives that are measurable. In addition, since measurable objectives can help clarify goals, studying assessment tools may cause departments to reexamine their goals to ensure that they really are the competencies the faculty wants their students to have attained.

When thinking about assessment measures, what comes first to the minds of mathematicians are grades and student performance on class tests. Tracking student grades over a number of years can provide some useful information. However, as a stand-alone tool it is not particularly good at revealing the kind of information faculty need in order to improve their program, and, after all, that's a major part of what assessment is all about. The case study from Colorado School of Mines (p. 149) illustrates how to incorporate traditional classroom testing into a comprehensive assessment process.

Cases studies gathered prior to and during the SAUM project include a wide variety of examples of assessment instruments including surveys, portfolios [Columbia College (Hopkins, 1999) and Northern Illinois University (Sons, 1999)], senior exit surveys [Colorado School of Mines (p. 149)], interviews and focus groups [Mary Washington College (Sheckels, 1999)], alumni surveys [St. Peter's College (p. 183)], employer surveys [University of Arkansas, Little Rock (p. 201)], senior capstone courses [Saint Mary's College of Indiana (Peltier 1999)], senior comprehensive exams—both written and oral [Franklin College (Callon, 1999) and Wabash College (Gold, 1999)], independent research projects [South Dakota State University (p. 191)], the Educational Testing Service's Major Field Test in Mathematics [University of Arkansas, Little Rock (p. 201)]and written and oral presentations of mathematics [Keene State College (p. 157) and Saint Mary's University of Minnesota (p. 177)].

In some instances one particular assessment tool might be effective in measuring a specific learning objective, while for other objectives a combination of instruments might work best. One lesson learned by SAUM participants is not to overwhelm the assessment program with so much data that it is not feasible to interpret the results. Point Loma Nazarene University's report, "Keeping Assessment Simple" (p. 163) addresses just this issue. In addition to keeping it simple, I would add, manageable.

Having developed a list of instruments that are associated with measurable objectives, it is likely that some tools will serve more than one purpose. That makes it easier to whittle down the potential instruments to a manageable number. It is also likely that not all of the department's objectives for mathematics majors can be measured. This poses a dilemma: either consider different or additional assessment tools, or decide that certain objectives can be measured in different ways, or (once again) revise the objectives.

Implementation of an assessment program leads to yet more decisions. Once the data has been collected it needs to be analyzed. The first issue is *who* should do it. Will it be the department chair, a subcommittee, or the entire department? Each choice comes with some advantages and disadvantages. Just as important, if not more important, is *how* to make sense of all of the data that has been gathered. Here's where rubrics and benchmarks are helpful to ensure consistency and faithfulness to the purposes of the assessment program. Assessment data is gathered by use of one or more instruments that were chosen to measure certain student learning objectives associated with particular departmental goals. Hence the department needs to (a) develop a rubric by means of which individual student responses can be judged and (b) establish a benchmark against which the composite of all those individual judgments can be measured. Case studies that describe this process include those from American University (p. 143), Colorado School of Mines (p. 149), Columbia College (Hopkins, 1999), and Northern Illinois University (Sons, 1999).

The final stage is the feedback loop. What do the results tell the department about whether students are meeting the goals the faculty has set forth and whether the program for mathematics majors is designed in such a way that the students can meet those goals? In other words, if students are falling short, is it the students who are underperforming or is it the program that is deficient? Whichever it is, in the final analysis the faculty must figure out how to improve the department's program so that mathematics majors are successful in developing the competencies the faculty desires for their majors. Sometimes (rarely) nothing needs to be done; sometimes the assessment plan needs to be fine-tuned; sometimes small changes in the major are sufficient; and sometimes the assessment results suggest significant curricular issues that need to be dealt with. After all, assessment is all about making the major better for our students. Examples of the variety of responses to assessment results can be seen in the case studies from American University (p. 143), Colorado School of Mines (p. 149), Columbia College (Hopkins 1999), Mary Washington College (Sheckels 1999), Point Loma Nazarene University (p. 163), and Wabash College (Gold 1999).

If developing a full-blown assessment plan seems daunting (as it did to several SAUM participants), one can proceed in stages. For example, once goals and learning objectives are established, it may be prudent to choose just one goal and its learning objectives to guide and pilot test the assessment process. Then assessment tools can be built to measure these objectives only. Analyzing these results, refining the plan, and making recommendations to improve the mathematics program in this one area can serve as a prototype for more comprehensive assessment in the future. Sometimes it is easier to see if one is on the right track by taking smaller steps, and often these lead to a better overall plan. The smaller-steps approach is discussed in case studies from Keene State College (p. 157), Point Loma Nazarene University (p. 163), Portland State University (p. 171), Saint Mary's University of Minnesota (p. 177), and Washburn University (p. 213).

One case study addresses a serious problem that is not altogether uncommon: what to do if assessment is mandated but the department faculty do not see any useful purpose in developing a plan? Such was the case at the University of Nevada, Reno (p. 207). Since the department chair is ultimately responsible, he or she can (and usually must) proceed alone. One approach is for the chair to assess a part of the curriculum that everyone agrees is problematic. The results of the analysis might lead to some good suggestions for improvement. In itself this might convince others to see the efficacy of assessment. But even if it doesn't, it will solve one problem that needs resolution.

Each undergraduate mathematics program is unique, yet there is enough commonality that we can learn from others' experiences. The case studies presented in this volume and its predecessor (Gold, et al., 1999) offer considerable variety of approaches to assessing the mathematics major. Even though none will be a perfect match for any other department, all offer ideas worth considering as departments plan to develop their own assessment programs.

References

Callon, G. Daniel. "A Joint Written Comprehensive Examination to Assess Mathematical Processes and Lifetime Metaskills." In *Assessment Practices in Undergraduate Mathematics*, Bonnie Gold, et al., eds. Washington DC, Mathematical Association of America, 1999, pp. 42–45. www.maa.org/saum/maanotes49/42.html

Committee on the Undergraduate Program in Mathematics. "Assessment of Student Learning for Improving the Undergraduate Major in Mathematics," *Focus: The Newsletter of the Mathematical Association of America*, 15(3) (June, 1995) pp. 24–28. Reprinted in *Assessment Practices in Undergraduate Mathematics*, Bonnie Gold, et al., eds. Washington, DC: Mathematical Association of America, 1999, p. 279–284. www.maa.org/saum/maanotes49/279.html

Gold, Bonnie, Sandra Z. Keith, and William A. Marion. *Assessment Practices in Undergraduate Mathematics*. MAA Notes No. 49. Washington, DC: Mathematical Association of America, 1999. www.maa.org/saum/maanotes49/index.html

Gold, Bonnie. "Assessing the Major Via a Comprehensive Senior Examination," In *Assessment Practices in Undergraduate Mathematics*, Bonnie Gold, et al., eds. Washington DC, Mathematical Association of America, 1999, pp. 39–41. www.maa.org/saum/maanotes49/39.html

Hopkins, Laurie "Assessing a Major in Mathematics." In *Assessment Practices in Undergraduate Mathematics*, Bonnie Gold, et al., eds. Washington DC, Mathematical Association of America, 1999, pp. 21–23. www.maa.org/saum/maanotes49/21.html

Peltier, Charles "An Assessment Program Built Around a Capstone Course." In *Assessment Practices in Undergraduate Mathematics*, Bonnie Gold, et al., eds. Washington DC, Mathematical Association of America, 1999, pp. 27–30. www.maa.org/saum/maanotes49/27.html

Sheckels, Marie P. "The Use of Focus Groups Within a Cyclic Assessment Program." In *Assessment Practices in Undergraduate Mathematics*, Bonnie Gold, et al., eds. Washington DC, Mathematical Association of America, 1999, pp. 35–38. www.maa.org/saum/maanotes49/35.html

Sons Linda, "Portfolio Assessment of the Major." In *Assessment Practices in Undergraduate Mathematics*, Bonnie Gold, et al., eds. Washington DC, Mathematical Association of America, 1999, pp. 24–26. www.maa.org/saum/maanotes49/24.html

Learning Outcomes Assessment: Stimulating Faculty Involvement Rather Than Dismay

Dan Kalman, Matt Pascal,
Virginia Stallings
Department of Mathematics and Statistics
American University
Washington, DC
kalman@american.edu,
mp3265a@american.edu,
vstalli@american.edu

Abstract. While academic institutions recognize the importance of implementing plans for assessment of learning outcomes of undergraduate programs, getting faculty to buy into the process can be difficult. This article will outline a university's attempts to engage faculty in learning outcomes assessment and how the Department of Mathematics and Statistics is beginning to incorporate what they have learned into a cycle of assessment that will inform their programs.

Introduction

American University (AU) is a private liberal arts college located in Washington, DC. It has approximately 10,000 students, 60% of whom are undergraduates. The Department of Mathematics and Statistics has 15 full-time faculty and 8 part-time instructors who support the university's mathematics competency requirement. Their programs include a bachelors and masters in both mathematics and statistics. The department currently has 15 undergraduate majors, mostly in the mathematics program. In a shift to support larger, self-sustaining programs, the university terminated the smaller PhD programs in Statistics and in Mathematics Education. The Mathematics and Statistics faculty turned their attention to revitalizing the undergraduate major programs which are showing promise through increasing interest and enrollments. They hope that focusing on assessment of learning outcomes will contribute to their ability to improve their programs and expand their numbers.

Background

In summers of 2001 and 2002, a small group of AU faculty and staff were sent to the American Association of Higher Education's Assessment conference to gather information on implementing an assessment process as a routine part of the university's annual review. When the group returned and reported their findings to the Provost, they decided to start a process of assessment by focusing on undergraduate major programs. This effort was initiated in the fall of 2001 by charging departments with writing their program goals and learning outcomes using an internally designed learning goals and objectives form. The Department of Mathematics and Statistics received the form from the Provost's Office (Appendix A) and proceeded to have discussions on how to complete the form for its undergraduate programs in Mathematics, Applied Mathematics, and Statistics. By Spring of 2002, the department submitted their Learning Goals for the BS in Mathematics (Appendix B, first section).

Unfortunately, there was very little information on how to complete the forms, and while the department considered very carefully what it valued and taught, their submission of learning outcomes reflected the traditional gauge of assessing programs through grades in courses, teaching evaluations, and vague assessments of outcomes. Departments across the university completed the forms in a similar fashion and then turned their attention to pressing issues raised by the university president's "Fifteen Points Plan" which included shifting the university's focus from graduate to undergraduate education, restructuring the faculty senate, and reducing the overall size of the university. Thus, the Mathematics and Statistics faculty were immersed in an intensive PhD program review that would ultimately lead to the elimination of their doctoral programs.

By spring 2003, the PhD program review was completed and the new streamlined faculty senate was formed. A substantial portion of the Middle States self-study was underway, which led to the creation of a university faculty and staff Project Team on Learning Outcomes and Assessment charged with examining the progress of the earlier learning outcomes submissions. Upon reviewing the assessment plans, the team found that additional action was needed to help departments generate active assessment cycles that truly informed programmatic change. The Project Team sponsored workshops that gave faculty hands-on experience in writing learning outcomes for their programs using materials developed by the faculty team. The materials included recommendations for improving assessment plans and encouraged departments to write learning objectives that focus on *results* instead of *process*. Departments were charged with identifying three to six learning outcomes that are critical to the program and are *observable* and *explicit*. The faculty team emphasized listing learning objectives separately, especially if they would require separate and different assessment strategies. They urged departments to visit their professional web sites and go to discipline-specific assessment conferences.

The Mathematics and Statistics Department's course of action

After attending the Project Team's workshops and reading their materials, the department decided to send a team to the Mathematical Association of America's *Supporting Assessment in Undergraduate Mathematics* (SAUM) series of three workshops offered over a three-year period. During the initial workshop in Highpoint, North Carolina, the team drafted a sample revised program to present to the department. The goal of the team was to present the department with a concrete example of a learning outcome and possible ways to assess the outcome (Appendix B, second section). They hoped that this example would provide a basis for comparing and improving the original assessment plans.

The team members purposefully included multiple forms of assessment to serve as a springboard for discussion when they presented the revised goal to the department. When they met with the department, they compared the original goals of the department with the revised simplified (illustrative) goal.

The team suggested that the department build its learning objectives from a program goal which the team drafted from the department's mission statement. The discussion was spirited and sometimes negative. Faculty were not convinced that this process was useful and they were skeptical that there would be meaningful results. Several faculty members opined that information for improving the programs could be obtained more simply by reflecting on their students' performances and discussing their observations about students among one another. One faculty member was insistent that since the program was so small, faculty *knew* their students in terms of their strengths and weaknesses.

Members of the team countered that while informal discussions are helpful, they do not pinpoint precisely the skills that are lacking (or are prevalent) in majors. The team emphasized the importance of instituting a process that produces observable evidence of student learning which in turn informs program improvement. The team suggested that if the department reviewed students' attempts at proofs, there would be variability in the degree of rigor and in the soundness of presentation. Setting up a situation that would allow the department to observe students' work would be relatively straight-forward as long as there was a specific targeted outcome, such as appropriate use of notation and terminology in the conduct of a proof. The team convinced faculty that assessment does not have to be drawn out or complicated and — in fact — can be conducted with processes that already exist. In short, what do we have to lose by taking a focused look at our students' behaviors when asked to perform a task that reflects skills gained through pursuing our program?

A turning point in the discussion came when one of the team members asked the mathematics faculty, "When our students graduate, what activity that is observable do you think our majors should be able to do?" A faculty member responded that he would like students to be able to pick up a mathematics book and teach themselves some new mathematics using the skills they have gained in the program. Other faculty members agreed and then the discussion shifted to developing a rubric that would describe the level of performance expected from a mathematics major. Ultimately, the department assigned a mathematician and a statistician to write the first learning outcome and assessment for their respective programs (Appendix B, third and fourth sections).

What's next

The faculty plan to develop at least two or three more learning outcomes to be included in an on-going assessment cycle. Rubrics need to be constructed for the mathematics self-teaching experience. The assessments for each of the above learning outcomes will begin fall semester 2004. The results will be analyzed in May of 2005. At that time, faculty will review the report on the analyses and then make decisions on what programmatic changes (if any) need to be made to respond to the results. In the meantime, the cycle for assessing other outcomes will begin in fall semester 2005 along with reassessing the first set of outcomes.

It would be misleading to say that all faculty in the department are confident that using learning outcomes to assess their programs will provide information that will improve the program. However, shortly after the assessment meeting, students in the graduate program in statistics gave oral presentations on their internships. When faculty were viewing the presentations, they began to discuss the poten- tial for including these presentations as part of the assess- ment cycle. They noted the strengths and weaknesses of the presentations and asked students for feedback on the pro- gram. A positive outcome was that students needed and asked for more writing activities in their courses. It was an *observable* result that showed faculty the efficacy of learn- ing outcomes assessment.

Bibliography

Undergraduate Programs and Courses in the Mathematical Sciences: CUPM Curriculum Guide 2004. Washington, DC, Mathematical Association of America.

Middle States Association of Colleges and Schools. Commission on Higher Education. (2002). *Characteristics of Excellence in Higher Education : Eligibility Requirements and Standards for Accreditation.* Philadelphia, PA: Commission on Higher Education, Middle States Association of Colleges and Schools.

Appendix A. Institutional Designed Assessment Form for American University

ACADEMIC OUTCOMES

Department:

Program:

Program Goals:

Expected Student Objective/Outcomes:

The undergraduate degree in [   ] emphasizes knowledge and awareness of:

Objective/Outcome 1:

Objective/Outcome 2:

Methods of Assessment	Standard for Success	Results	Action/Steps/Comments

Methods of Assessment	Standard for Success	Results	Action/Steps/Comments

Appendix B. Learning Outcomes for the B. S. in Mathematics

Original Submission, Spring 2002

Learning Goal I. Students are expected to acquire ability and skills in Calculus of one and several variables, vector analysis, basic linear algebra and elements of the theory of vector spaces. Students are to develop appreciation for mathematical reasoning and acquire skills in logical deduction. Ability to formulate definitions, to apply the methods of direct proof and indirect proof to solve problems is expected on a basic level. Emphasis is given to developing ability to communicate effectively in explaining the overall processes and the particular steps in the solving of a mathematical problem.

Learning Goal II. The mathematics major is expected to develop a fundamental understanding of several major realms of mathematics. Students are expected to understand the different methods in real analysis and modern algebra, and be able to apply the methods in a rigorous manner. Further understanding of the span of mathematics is expected in the curriculum of a major. Students should be able to demonstrate understanding of the basic methods of inquiry in at least three specific areas in mathematics. Areas of expertise of faculty members include history of mathematics, mathematical logic, set theory, complex analysis, differential equations, geometry, number theory, topology, harmonic analysis, numerical analysis, probability and statistics. Ability to communicate mathematical ideas clearly and logically is given continued emphasis.

The scope and depth of the program provide students with the ability to continue their study in a graduate program, or to teach in classrooms, or to enter the industrial world.

Objective/Outcome 1: Ability and skills in Calculus of one and several variables, vector analysis, basic linear algebra and elements of the theory of vector spaces. Appreciation for mathematical reasoning and acquire skills in logical deduction. Ability to formulate definitions, to apply the methods of direct proof and indirect proof to solve problems is expected on a basic level. Ability to communicate effectively in explaining the overall processes and the particular steps in the solving of a mathematical problem.

Methods of Assessment. Evaluation of student's grades in the following courses:

　1. MATH-221 (Calculus I)　　3. MATH-223 (Calculus III)　　5. MATH-322 (Advanced Calculus)
　2. MATH-222 (Calculus II)　　4. MATH-310 (Linear Algebra)

Standards for Success. Each student receives a grade of C or better in each of these five courses

Objective/Outcome 2: Understanding of the different methods in real analysis and modern algebra, and be able to apply the methods in a rigorous manner.

Methods of Assessment 1: Evaluation of student's grade in the following courses:

　MATH-512 (Intro to Modern Algebra I)　　　　MATH-520 (Intro to Analysis I)
　MATH-513 (Intro to Modern Algebra II)　　　　MATH-521 (Intro to Analysis II)

Standards for Success. Each student receives a grade of C or better in each of these four courses.

Methods of Assessment 2: Supplemental questions specifically addressing this objective on student Course Evaluations for MATH-513 and MATH-521.

Standards for Success: Majority of students "strongly agree" or "agree" to each of the supplemental questions.

Revised Draft for Departmental Consideration (Spring 2004)

Program Goals: Our goals include teaching the essential skills of mathematical literacy and proficiency. Literacy and proficiency in mathematics include not only the ability to comprehend mathematical reasoning but also the ability to express oneself mathematically: to formulate an argument as well as follow it. Our students will be able to understand and apply mathematics as a model for finding solutions to real-life problems.

Objective/Outcome 1: Student will be able to apply the methods of direct proof and indirect proof to solve problems.

Methods of Assessment:

- Final examinations from Advanced Calculus, Linear Algebra, Analysis I, Analysis II, Modern Algebra I and Modern Algebra II will each have a department selected problem that requires skill in use of direct or indirect proofs.
- Give an exit exam that contains solving problems using direct and indirect proofs. (For assessment of program)

- Exit interview to include question on how to solve a problem using a direct proof or indirect proof.
- Folder on each major's finals
- Portfolio

Standard for Success: Faculty will develop a rubric that describes the expected characteristics of solving a problem using direct or indirect proof.

Objective/Outcome 2: Students will be able to develop a mathematical model from a real life application.

Methods of Assessment: Give an exit exam that contains at least one posed application problem.

Standard for Success: Faculty will develop a rubric that describes the expected characteristics of modeling.

Revised Outcomes for B.S. in Mathematics, Summer 2004

Mathematics, B. S. Program Goals: Our goals include teaching the essential skills of mathematical literacy and proficiency. Literacy and proficiency in mathematics include not only the ability to comprehend mathematical reasoning but also the ability to express oneself mathematically: to formulate an argument as well as follow it. Our students will be able to understand and apply mathematics as a model for finding solutions to real-life problems.

Mathematics Objective/Outcome 1: Student will be able to orally explain a concept in mathematics from an advanced level mathematics text (the concept should be one that immediately or closely follows the last concept discussed in one of the last courses in their program).

Methods of Assessment: Seniors will give an oral/chalkboard presentation to three faculty members.

Standards for Success: Faculty will develop a rubric that describes the expected characteristics of self-teaching a concept including proper use of notation and procedures for interpreting and explaining a mathematical concept.

Revised Outcomes for B.S. in Statistics, Summer 2004

Statistics, B. S. Program Goals: Our goals include teaching the essential skills of statistical literacy and proficiency. Literacy and proficiency in statistics include not only the ability to comprehend statistical reasoning but also the ability to use and interpret data effectively. Our students will be able to understand and apply statistics as a model for finding solutions to real-life problems.

Statistics Objective/Outcome 1: The student will be able to summarize and describe data, conduct graphical analyses, carry out basic formal statistical procedures and effectively write up the analysis.

Methods of Assessment: In each of the statistical methods courses (STT515, STT516, STT521, STT522, STT424) professors will assign at least one project in which students use the methods learned in that class to explore and analyze a complex data set and write up the analysis. Professors will keep a copy of the projects of the statistics majors for the purposes of assessing the outcome stated above.

Standards for Success: At the end of each academic year a team of three statistics faculty will evaluate the project according to an agreed upon rubric.

Evaluation of the projects: At the end of each academic year a team of three statistics faculty will evaluate the project according to an agreed upon rubric:

1. Concise but clear description of problem.
2. Description of method used for analysis, including a discussion of advantages, disadvantages and necessary assumptions.
3. Discussion of results.
4. Conclusion including a discussion of limitations of analysis.

Standards:

a. Advanced: Easy to read, concise correct with all important pieces of information included. Appropriate use, display and description of graphs.
b. Proficient: Correct statements with all important pieces of information included. Appropriate use of graphs.
c. Basic: Correct statements, but some important aspects of the problem omitted.
d. Unacceptable: Incorrect statements. Inappropriate use of graphs. Unintelligible sentences.

Expected Standard for the program: A majority of students will score at least proficient on all four pieces of the rubric.

The Development, Implementation and Revision of a Departmental Assessment Plan

Barbara M. Moskal
*Mathematical and Computer Sciences
Department
Colorado School of Mines
Golden, CO*
bmoskal@mines.edu

Abstract. In the late 1990s, the Colorado School of Mines began to prepare for a visit by the Accreditation Board for Engineering and Technology. As a support department to eight accredited engineering departments, the Mathematical and Computer Sciences Department had a responsibility to assist in the accreditation process. The approaching visit motivated the creation of a departmental assessment plan. This paper traces the development and implementation of the original departmental assessment plan and the events that stimulated the revision of that plan. An emphasis will be placed throughout this paper upon what the Mathematical and Computer Sciences department has learned.

Background

In the late 1990s, the Colorado School of Mines (CSM) began to prepare for a visit by the Accreditation Board for Engineering and Technology (ABET) that was to take place in the academic year 2000–2001. Eight CSM engineering departments were to be reviewed using ABET's newly revised engineering criteria [1]. A major difference between the old and the new criteria was that the new criteria required that accredited departments directly demonstrate what students know and can do. In other words, the new criteria emphasized the direct assessment of student outcomes.

All of the engineering departments and the departments that support core-engineering courses worked to develop and implement an assessment system that measured student outcomes. As a department that provides many of the core engineering courses, the Mathematical and Computer Sciences Department (MCS) began to develop an assessment plan in the fall of 1997. The first step in this process was to establish departmental goals and objectives. This was completed by the end of the 1997–98 academic year. The next two years were dedicated to developing and implementing a departmental assessment plan that would support the student attainment of the departmental goals and objectives. Much of the original assessment plan focused upon how the information that was currently being collected could be better used for assessment purposes. A number of survey instruments were also introduced with the purpose of acquiring data in a fast and efficient manner. In summary, much of the early work with respect to assessment process was focused upon ensuring a successful ABET visit.

The ABET visit in 2000–2001 resulted in all eight accredited departments receiving full accreditation. The success of this visit immediately reduced the administrative pressure to produce quick assessment results. The MCS department now had the time and opportunity to review and revise the department's assessment plan. The review process began in the fall of 2001. By the end of the academic year, the department's Undergraduate Curriculum Committee had approved a revised set of goals and objectives and a revised departmental assessment plan. The MCS department is currently in the process of implementing the revised plan.

This paper traces the development and implementation of the original MCS departmental assessment plan and the events that stimulated the need to revise that plan. An emphasis will be placed throughout this paper upon what the MCS department has learned.

Conceptual framework

According to the Mathematical Association of America's (MAA) Committee on the Undergraduate Program in Mathematics (CUPM) in collaboration with the MAA's Assessment Subcommittee, assessment is a cycle that consists of the following five phases [2]: 1) articulating goals and objectives, 2) developing strategies for reaching goals and objectives, 3) selecting instruments to evaluate the attainment of goals and objectives, 4) gathering, analyzing and interpreting data to determine the extent to which goals and objectives have been reached, and 5) using the results of assessment for program improvement. When the final phase is reached, the assessment cycle begins again. This conceptualization of the assessment process is consistent with other literature on assessment [3, 4, 5]. The phases within this cycle provide a framework for developing a departmental assessment plan. Each phase can be moved through sequentially, supporting the emergence of a departmental assessment plan. Once the initial cycle has been completed, the knowledge and information that has been gained through the implementation process can be used to improve the assessment plan prior to the next cycle.

Development and revision of goals and objectives

As was discussed earlier, the first step in the development of a departmental assessment plan is the establishment of goals and objectives. Goals are broad statements of expected student outcomes and "objectives" divide a goal into circumstances that suggest whether a given goal has been reached [6]. Careful thought should be given to the University Mission Statement [2, 7] and to the requirements of any appropriate accreditation board when developing departmental goals and objectives [7]. Attention should also be given to faculty buy-in. Faculty will not work to assist students in reaching goals and objectives that they do not believe are important. Departmental goals and objectives

Original Statement*	Revised Statement*
G1: Develop technical expertise within mathematics/computer science	G1: Students will demonstrate technical expertise within mathematics/computer science by:
O1: Design and implement solutions to practical problems in science and engineering	O1: Designing and implementing solutions to practical problems in science and engineering,
O2: Use appropriate technology as a tool to solve problems in mathematics/computer science	O2: Using appropriate technology as a tool to solve problems in mathematics/computer science, and
O3: Create efficient algorithms and well structured programs	O3: Creating efficient algorithms and well structured computer programs.
G2: Develop breadth and depth of knowledge within mathematics/computer science	G2: Students will demonstrate a breadth and depth of knowledge within mathematics/computer science by:
O4: Extend course material to solve original problems	O4: Extending course material to solve original problems,
O5: Apply knowledge of mathematics/computer science	O5: Applying knowledge of mathematics/computer science to the solution of problems,
O6: Identify, formulate and solve mathematics/computer science problems	O6: Identifying, formulating and solving mathematics/computer science problems, and
O7: Analyze and interpret data	O7: Analyzing and interpreting statistical data.
G3: Develop an understanding and appreciation for the relationship of mathematics/computer science to other fields	G3: Students will demonstrate an understanding and appreciation for the relationship of mathematics/computer science to other fields by:
O8: Apply mathematics/computer science to solve problems in other fields	O8: Applying mathematics/computer science to solve problems in other fields,
O9: Work cooperatively in multi-disciplinary teams	O9: Working in cooperative multi-disciplinary teams, and
O10: Choose appropriate technology to solve problems in other disciplines	O10: Choosing appropriate technology to solve problems in other disciplines.
G4: Communicate mathematics/computer science effectively	G4: Students will demonstrate an ability to communicate mathematics/computer science effectively by:
O11: Communicate orally	O11: Giving oral presentations,
O12: Communicate in writing	O12: Completing written explanations,
O13: Work cooperatively in teams	O13: Interacting effectively in cooperative teams,
O14: Create well documented programs	O14: Creating well documented programs, and
O15: Understand and interpret written material in mathematics/computer science	O15: Understanding and interpreting written material in mathematics/computer science.
*G: Goals, O: Objectives	

Table 1. General Statement of Student Goals and Objectives.

should reflect the collective understanding of the faculty members of what students should know and be able to do, rather than the ideas of a single individual.

Departments may be tempted to use the current curriculum to motivate the development of student goals and objectives. Although this method will result in the perfect alignment of the goals and objectives with the curriculum, it will also result in a missed opportunity for improving the curriculum. In other words, the process of determining what is important should not be artificially constrained by what currently exists.

In order to support the development of a set of goals and objectives and to acquire faculty support for the goals and objectives, each member of the full-time MCS faculty was interviewed in the fall of 1997. They were asked:

1. What competencies do you think students should have after completing the mathematics core courses?,
2. What competencies do you think students should have after completing their major courses in mathematics?, and
3. What competencies do you think students should have after completing their major courses in computer science?

The reader will notice that each of these questions refers to student competencies rather than student goals and objectives. This phrasing stimulated the faculty to identify specific knowledge and skills. A feature of the interview process was that the faculty were not directed to consider the current curriculum, but rather they were asked to indicate the competencies that students should have upon completion of their course work.

Using the specific information that the faculty provided, the departmental assessment specialist created broader statements of goals that captured these competencies. Next, a departmental sub-committee was formed that consisted of the head of the department, a mathematician, a computer scientist, a mathematics education expert and an assessment specialist. Based on the faculty responses to the interview process, the requirements of ABET and the University Mission Statement, the sub-committee drafted four sets of departmental goals and objectives: 1) a general statement for all students of mathematics and computer science, 2) a statement for the core mathematics courses, 3) a statement for the major mathematics courses, and 4) a statement for the computer science major courses. The fulltime faculty approved a version of these goals and objectives later that academic year.

Periodically, the established departmental goals and objectives should be reviewed to determine whether they continue to be consistent with departmental needs. In the academic year 2001–2002, the department's Undergraduate Curriculum Committee reviewed the student goals and objectives. Although the committee felt that this list continued to capture the desired student outcomes, questions were raised with respect to the phrasing of the goals and objectives. The original list consisted of short phrases that implied the desired student outcome. The revised list directly indicated what the students needed to demonstrate in order to suggest that a given goal had been reached. Both the original and the revised list of student goals and objectives are displayed in Table 1.

A critical question was raised during the final review of the revised student goals and objectives, "What is the faculty's responsibility in assisting students in reaching these goals and objectives?" This question resulted in the development of the faculty goals and objectives that are shown in Table 2.

G1: Faculty will demonstrate technical expertise within mathematics/computer science by:
  O1: Providing clear, technical explanations of mathematics/computer science concepts to students,
  O2: Using appropriate technology as a tool to illustrate to students how to solve mathematics/computer science problems, and
  O3: Providing examples of how mathematics/computer science can be applied to the solution of problems in other fields.
G2: Faculty will support the students attainment of the goals and objectives outlined above by providing the students the opportunity to:
  O4: Solve original problems, some of which are drawn from other fields,
  O5: Use technology as a tool in solution of mathematics/computer science problems,
  O6: Design algorithms and structured programs,
  O7: Identify, formulate and solve mathematics/computer science problems,
  O8: Interact in cooperative teams,
  O9: Give oral presentations,
  O10: Communicate in writing, and
  O11: Interpret written material in mathematics/computer science.
G3: Faculty will evaluate the students attainment of the above goals and objectives outlined above by creating assessments for the evaluations of students ability to:
  O12: Solve original problems, some of which are drawn from other fields,
  O13: Use technology as a tool in solution of mathematics/computer science problems,
  O14: Design algorithms and well structured programs,
  O15: Identify, formulate and solve mathematics/computer science problems,
  O16: Interact in cooperative teams,
  O17: Give oral presentations,
  O18: Communicate in writing, and
  O19: Interpret written material in mathematics/computer science.

G: Goals, O: Objectives

Table 2. Faculty Goals and Objectives

The faculty goals and objectives were designed to parallel the student goals and objectives, indicating the faculty's responsibility in supporting the desired student outcomes.

Using goals and objectives to develop, implement and revise an assessment plan

Once a program has a set of goals and objectives, the next step is the creation of a plan that will support the attainment of those goals and objectives. In order to facilitate the creation of this plan, the MCS department used the Olds and Miller Assessment Matrix [8, 9]. This matrix, which is available electronically,[1] provides a framework for planning and recording the phases of the assessment cycle. An example of a portion of the current student assessment plan is shown in Table 3. In the academic year 2001–2002, this plan was extended to include a faculty assessment plan. Both components of the larger assessment plan, student and faculty, are also available on-line.[2]

The first column, "Performance Criteria (PC)", is a statement of an observable performance that is used to determine whether a given objective has been reached. The statements within this column describe the strategies that will be used to interpret the collected data. Stating performance criteria is a first step in the development of strategies for reaching and evaluating the attainment of goals and objectives (phase 2 of the assessment process). The reader will notice that many of the established performance criteria in the MCS assessment plan reference a specific course. For example, PC1 states, "Students in Calculus for Scientists and Engineers (CSE) I, II and III will complete common exams that assess this objective. All students will pass the calculus sequence prior to graduation." The calculus sequence consists of three courses, all of which are coordinated. Coordinated courses have multiple sections, which are taught by different instructors. A lead faculty member coordinates these sections and holds regular meetings at which instructors have the opportunity to share instructional strategies and to create common assignments and/or exams. The lead faculty member for coordinated courses also ensures that the designated program objectives are assessed through common assignments and/or exams.

The MAA [2] has criticized the practice of many mathematics departments of restricting the assessment process to traditional testing techniques, which are characterized by examining individual student responses to written evaluations. Instead, the MAA has recommended that departments

---

[1] www.mines.edu/fs_home/rlmiller/matrix.htm

[2] www.mines.edu/Academic/assess/Plan.html

supplement information acquired through traditional methods with additional information that has been acquired through team assignments and presentations. One aspect of the MCS Performance Criteria is that it does not rely solely upon traditional testing methods. Several of the performance criteria within the MCS plan refer to common student assignments that require group work and/or oral presentations. For example, PC24 states, "All students are required to pass Engineering Practices Introductory Course Sequence prior to graduation. Successful completion of this course requires that students work in multidisciplinary teams for a semester on the solution of a problem that was solicited from a local business." Other papers have addressed how these team activities are evaluated [10, 11]. Performance Criteria has also been established for the evaluation of oral presentations, "PC27: Students complete team oral presentations in Field Session. All MCS majors are required to pass this course prior to graduation". Field Session is an intensive six week summer course that is completed immediately following the students' junior year. Students and their teams dedicate at least eight hours each day to solving a problem that has been solicited from a local business. At the conclusion of the course, the students are required to present their solution to the participating company and to their instructors in both written and oral form. These activities are then evaluated using a common scoring rubric or scoring scheme. Although PC27 is specific to team oral presentations, other criteria have been established that address individual oral presentations.

The second column in the assessment matrix is "Implementation Strategy." This refers to the student or faculty activities that support the attainment of given performance criteria. Stating which activities will support the attainment of the goals and objectives is the second step in developing strategies for reaching the goals and objectives (phase 2 of the assessment cycle). For the majority of the student goals and objectives in the MCS assessment plan, the implementation strategy is the students' coursework. This includes core courses (courses that are required of all CSM graduates), major courses (courses that are required of all MCS majors) and Field Session (a design course that is required of all MCS majors after their junior year).

As was discussed earlier, core courses are coordinated. A lead faculty member schedules regular meetings throughout the semester among the course instructors. During these meetings, instructors have the opportunity to share the activities that they have developed to support the attainment of the appropriate goals and objectives. In the major courses, it is the responsibility of the individual faculty member to ensure that the goals and objectives are being reached

	(1)	(2)	(3)	(4)	(5)
G1: Students will demonstrate technical expertise within mathematics/computer science by:					
Objectives (O)	**Performance Criteria (PC)**	**Implementation Strategy**	**Evaluation Method (EM)**	**Timeline (TL)**	**Feedback (FB)**
O1: Designing and implementing solutions to practical problems in science and engineering.	PC1: Students in Calculus for Scientists and Engineers (CSE) I, II and III will complete common exams that assess this objective. All students will pass the calculus sequence prior to graduation. PC2: Students in Programming Concepts and Data Structures will learn to use computer programs to solve problems. All majors in MCS will pass these courses prior to graduation. PC3: All MCS majors will pass field session prior to graduation. Field session requires that the student apply mathematics/computer science to the solution of original complex problems in the field. PC4: At least 80% of graduating seniors will agree with the statement, "My MCS degree prepared me well to solve problems that I am likely to encounter at work".	Core Coursework Major Coursework Field Session	EM1: PC1 will be evaluated by instructors of the calculus sequence. EM2: PC2 will be evaluated by instructors of Programming Concepts and Data Structures. EM3: PC3 will be evaluated by the Field Session instructors. EM4: PC4 will be evaluated through the senior survey.	TL1: EM1 implemented in F'97. TL 2: EM2 implemented in F'97 TL3: EM3 implemented in F'97 TL4: EM4 implemented in S'99	FB1: Verbal reports will be given to the under-graduate committee and the department head concerning student achievements within the respective courses at the end of each semester. FB2: Degree audit completed prior to graduation to ensure that all students completed requirements of degree. FB3: A written summary of the results of the senior survey will be given to the department head.

Table 3. Portion of the Mathematical and Computer Sciences Department's Student Assessment Plan[3]

within a given course. The importance of reaching the stated goals and objectives through these courses is further reinforced in the faculty assessment plan in that faculty are expected to provide students with the opportunity to attain each of the student goals and objectives.

The third column in the assessment matrix, "Evaluation Methods," specifies the measurement instrument that will be used to collect the evidence as to whether the performance criteria have been reached. This column describes phase three of the assessment cycle (i.e., selecting assessment instruments). Two types of evaluations are mentioned in the student assessment plan: 1) common assignments and/or exams in coordinated courses and 2) the student senior survey. Three methods are referenced in the faculty assessment plan: 1) student evaluations of faculty instruc-

tional efforts, 2) voluntary observations completed during classroom instruction, and 3) a review of course materials by the course coordinator.

The majority of the evaluation methods that were used in the original student assessment plan were dependent upon surveying the faculty and students. Although survey instruments provide an easy manner in which to acquire information, it is not necessarily the most reliable method of data collection. When the plan was reviewed and revised in 2001–2002, the decision was made to shift the focus to measurement techniques that directly assess student performances. This resulted in the current student assessment plan, which is dependent upon student performances on assignments, exams, oral presentations and team activities that are completed in classes. Surveys have not been completely eliminated from the revised plan. Students, upon graduation, are still asked to complete a survey and this information is used to supplement the direct measurement

[3] The complete plan is online at www.mines.edu/Academic/assess/ Plan.html.

Semester	Source	Concern	Response	Follow-up
Spring '00	Course evaluations	A set of open-ended questions were added to the course evaluations in 1997. The average faculty rating on each question in 1997 was compared to the average faculty rating in 2000. The faculty ratings have increased since the changes have been implemented.	Current faculty evaluation system will be maintained.	Open-ended questions continue to be used as part of the faculty evaluation.
Spring '00	Senior Survey	The senior survey indicated that many of the graduating seniors felt that the had inadequate skills in written communication.	Acting Department Head and Coordinator for Probability and Statistics course attended a summer workshop on how to introduce writing in summer workshops.	Writing assignments have been added to the Probability and Statistics course.
Fall '00	Course evaluations	Concern was raised that faculty do not use the information that is provided by students in response to the faculty evaluations.	A set of questions was developed that asks faculty to examine the student evaluations and to write a response as to how they would use the information to improve the course.	Faculty continue to respond in writing to the student evaluations each semester. Department head reviews the response.
Spring '01	Course evaluations Feedback from Engineering Division	Concern was raised about the content of the Probability and Statistics for Engineers Course.	A new book was selected that better meets the needs of engineers. Additionally, new labs were created with the same purpose in mind.	A mini-grant was sought and acquired to support the improvement of this course in an appropriate manner. As part of this effort, a survey has been developed and is administered each year concerning students' experiences in the course.
Fall '01	General Review of Assessment System	A general review of our assessment system indicated that a number of our instruments and methods are out of date.	An effort was begun to revise the goals, objectives and overall system in a manner that is appropriate to our current needs.	The revised goals, objectives and overall system were reviewed and approved by the Undergraduate Committee in the Spring '02.

Table 4. A Portion of the MCS Feedback Matrix[4]

of the student outcomes. Given that this is the first year of implementation for the revised plan, the new methodology has not yet been fully tested.

"Timeline" refers to when each evaluation method will be implemented. This column contains information on when the data will be gathered, analyzed and interpreted (phase four of the assessment cycle). Examination of the presented timeline suggests that the new assessment techniques have been introduced slowly during the five years in which the plan has existed. Spacing the introduction of new techniques has allowed the department to focus upon the implementation of specific methodology within a given year or semester. Had the department introduced all of the new measurements in the first year, the department would have been overwhelmed and the assessment plan would have failed.

The final column of the assessment matrix is "Feedback." This column indicates how the acquired information will be disseminated and used and is directly linked to the fifth phase of the assessment process (using results). A primary concern that has been expressed by the MAA [2] and others [1, 3, 4] with regard to assessment is ensuring that the information acquired through assessment is used for program improvement. At the end of each semester, the department's assessment specialist summarizes the results of all survey instruments and compares these outcomes to the stated performance criteria. She also meets with the coordinators of the core courses and discusses how the goals and objectives were supported and assessed. Based on this information, she then writes and submits a report to the head of the department. This report contains recommendations on how the department may improve its programs and its assessment system.

---

[4] Complete matrix online at: www.mines.edu/Academic/assess/Feedback.html

To assist the MCS department in documenting how the information that is acquired through assessment is used, a feedback matrix was developed [12]. A portion of this matrix is shown in Table 4; the complete feedback matrix can be found on line.[5] This matrix indicates in which semester the information was collected, the source of the information, the concern that the information raised, the department's response to that concern and the efforts to follow-up on whether the response was successful in addressing the concern. For example in the spring of 2001, the students in Probability and Statistics for Engineers indicated on the student evaluation that the content covered during the semester was not consistent with the statistics that they used in their engineering courses. Based on this, the MCS department asked the engineering division to review and provide feedback on the curriculum for the Probability and Statistics for Engineers course. The Engineering division made a number of suggestions, including the recommendation that error analysis become part of the curriculum. In the fall of 2001, a new textbook and lab manual was selected that was consistent with the Engineering divisions recommendations. This process is documented in the feedback matrix.

Conclusions

A great deal of time and effort has been dedicated to the development and implementation of the MCS departmental assessment plan. Revision of this plan is ongoing. As was discussed earlier, continual improvement of an assessment plan is a natural part of the assessment process. Many of the techniques that have been used here can be easily transported to the needs of other departments and disciplines. In fact, the Olds and Miller assessment matrix is already used across the CSM campus [12]. Further information concerning the MCS department's assessment efforts can be found at the department's assessment webpage,[6] together with additional information concerning the broader CSM assessment effort.[7]

Acknowledgements. The plan discussed here reflects the collaborative effort of the Mathematical and Computer Sciences Department (MCS) at the Colorado School of Mines (CSM). The author gives special recognition to Dr. Graeme Fairweather, MCS Department Head, and Dr. Barbara Bath, Director of Undergraduate Studies in MCS. Additionally, special thanks is given to Dr. Barbara Olds and Dr. Ron Miller who designed the assessment matrix that was used to organize the assessment plan presented here.

References

1. *Engineering Criteria,* Accreditation Board of Engineering and Technology, 6th ed., 2000, available at www.abet.org/accreditation/accreditation.htm.

2. *Assessment of Student Learning for Improving the Undergraduate Major in Mathematics,* Mathematical Association of America, Subcommittee on Assessment, Committee on Undergraduate Program Mathematics, 1995.

3. Steen, L, "Assessing Assessment,"in Gold, B., Keith, S.Z., and Marion, W., eds., *Assessment Practices in Undergraduate Mathematics,* 1999, pp. 1–6.

4. *Assessment Standards for School Mathematics,* National Council of Teachers of Mathematics (NCTM), Reston, Virginia, 1995.

5. Moskal, B. "An Assessment Model for the Mathematics Classroom," *Mathematics Teaching in the Middle School,* 6 (3), 2000, pp. 192–194.

6. Rogers, G. & Sando, J., *Stepping Ahead: An Assessment Plan Development Guide,* Rose-Hulman Institute of Technology, Terre Haute, Indiana, 1996.

7. Moskal, B. M. & Bath, B.B., "Developing a Departmental Assessment Plan: Issues and Concerns," *The Department Chair: A Newsletter for Academic Administrators,* 11 (1), 2000, pp. 23–25.

8. Olds, B. M. & Miller, R., "Assessing a Course or Project," *How Do You Measure Success? (Designing effective processes for assessing engineering education),* American Society for Engineering Education, 1996, pp. 135–44.

9. ———, "An Assessment Matrix for Evaluating Engineering Programs," *Journal of Engineering Education,* 87 (2), 1998, pp. 173–178.

10. Knecht, R., Moskal, B. & Pavelich, M., "The Design Report Rubric: Measuring and Tracking Growth Through Success," Proceedings of the annual meeting of the American Society for Engineering Education, St. Louis, Missouri, 2000.

11. Moskal, B., Knecht, R. & Pavelich, M., "The Design Report Rubric: Assessing the Impact of Program Design on the Learning Process," *Journal for the Art of Teaching: Assessment of Learning,* 8 (1), 2001, pp. 18–33.

12. Moskal, B., Olds, B. & Miller, R.L., "Scholarship in a University Assessment System," *Academic Exchange Quarterly,* 6 (1), 2002, pp.32–37.

---

[5] www.mines.edu/Academic/assess/Feedback.html

[6] www.mines.edu/Academic/assess/

[7] www.mines.edu/Academic/assess/Resource.htm

Assessing Student Oral Presentation of Mathematics

Dick Jardine and Vincent Ferlini
Department of Mathematics
Keene State College
Keene, NH
rjardine@keene.edu, vferlini@keene.edu

Abstract. Like many colleges and universities, the overall mathematics program at Keene State College (KSC) includes programs supporting the mathematics major, teacher preparation, developmental mathematics, general education, and service courses for other departments. For our initial assessment effort, we decided to focus on our major, and on one particular aspect of that program specifically. One of the department's goals is that our majors graduate with an ability to communicate mathematics effectively. We identified a specific learning outcome tied to that goal: that students demonstrate an ability to communicate mathematics effectively by giving oral presentations. This case study will address how we planned and implemented an assessment of that learning outcome in the fall semester of 2002, to include how we intend to use the assessment results to modify our program.

Background and Goals

KSC is a public liberal arts college, a Council of Public Liberal Arts Colleges (COPLAC) institution. We have 4,200 undergraduates, and we graduate 8 to 10 mathematics majors each year. The core of our mathematics major includes a statistics course, the first two courses of the traditional three-course calculus sequence, a transition course (Introduction to Abstract Math), and linear algebra. Students then choose an option (another eight or more courses) which focus on teacher preparation at the middle or secondary level, pure mathematics, applied mathematics, computer mathematics, and math-physics.

Our department established the following goals (based on our department mission in Figure 1) and provides the environment to enable students to accomplish those goals while a mathematics major at KSC. We expect that a KSC mathematics major will possess:

- Technical skill in completing mathematical processes;

- Breadth and depth of knowledge of mathematics;

- An understanding and appreciation of the relationship of mathematics to other disciplines;

- An ability to communicate mathematics effectively;

- A capability of understanding and interpreting written materials in mathematics;

- An ability to use technology to do mathematics.

In keeping with the mission of the college, the Mathematics Department of Keene State College provides and maintains a supportive intellectual environment that offers students mathematical experiences appropriate to their individual needs and chosen programs of study. The department provides an in-depth study of mathematics in preparation for either an immediate career, especially teaching, or graduate school; supports the mathematical needs of other academic disciplines; and maintains a program available to all students to enhance their ability to think mathematically and to reason quantitatively.

Figure 1. Department Mission

Our initial assessment effort focused on our graduates' ability to communicate mathematics effectively. A specific learning outcome tied to that goal is that our students demonstrate an ability to communicate mathematics effectively by giving oral presentations. There is consensus in the department that this is an important outcome, and as a department we chose to focus attention on that objective to begin the overall assessment of our major. Figure 2 provides an outline of the assessment process we implemented.

Learning outcome	Strategy to accomplish objective	Assessment plan	Data collection & interpretation	Program improvement
Students will demonstrate an ability to communicate mathematics effectively by a. giving oral presentations;	Student oral presentations part of undergraduate experience through many courses; Requirements and presentation expectations clearly communicated to the student; Students follow timeline leading to successful presentation; Students attend seminars to see examples of presentations by faculty and peers.	Rubric developed for faculty evaluation of seminar presentation. Rubric developed for student evaluation of seminar presentation. Student self-evaluation and interview conducted by instructor.	Completed rubrics compiled by course instructors. Instructors identification of strengths and weaknesses recorded.	Assessment subcommittee assembles information and reports to the department with recommendations for change, as needed. Department discusses implementation of recommended modifications, as needed.

Figure 2. Assessment Framework for Oral Presentation

Outcome

Many of our students begin making formal presentations early in our program; all of our majors make presentations later in the program. For example, students in an Introductory Statistics course make brief but formal PowerPoint presentations on group projects they have completed. One such project involved the use of descriptive statistics to compare populations of trout in local streams, based on data from the NH Fish and Game department. In those early courses, students are acquainted with the guidelines for making presentations and the rubrics instructors use to set the standards for student presentations. Our faculty use variations of the rubric in other mathematics courses that require presentations.

Students in most of our upper level mathematics courses make longer presentations of their project work, and some of those presentations are made not only before their instructor and peers, but also before department faculty as part of our weekly Friday Department Seminar. Additionally, our students have made presentations outside the department at our college-wide Academic Excellence Conference, at MAA Northeastern Section regional meetings, and at the Hudson River Undergraduate Mathematics Conference. As an example, a student in a recent history of mathematics course presented his project on fractal geometry in the course, at a department seminar, and at the college Academic Excellence Conference. By formalizing the assessment process, we have improved our students' abilities to make effective mathematical presentations through implementing a well-thought out strategy to ensure student success.

Description: What did we do?

We identified two courses for implementation of this initial assessment effort: MATH310 History of Mathematics and MATH463 Complex Variables. Students in those courses comprised a large percentage of our majors. The department agreed that students in those courses would make presentations at the end of the semester in the Department Seminar scheduled for the last week of classes. Faculty attending the seminar would evaluate the student presentations using the rubric (Appendix A) already in use. The results would be assembled by an assessment team and briefed to the department at the beginning of the spring semester, with recommendations offered by the team for improvement of our program and in the assessment process.

At the beginning of the fall semester, students in both courses, all mathematics majors, learned of the department mission and majors program goals. They understood that oral presentation skills were to be developed and assessed over the course of the semester, with evaluation of their presentations in the Department Seminar the last week of classes.

Several steps were taken to develop student presentation skills. Students were given a brief presentation on the effective use of PowerPoint in making presentations. In order to eliminate misunderstanding of the standards, their instructor explained the rubric to be used to evaluate their presentations, Students were encouraged to attend the weekly Department Seminars to observe faculty presentations. Over the course of the semester, students in both courses made several informal and formal presentations, some of which were assessed using the rubric, to increase their experience and comfort level with public speaking.

At the end of the semester, their instructor, other faculty members, and their peers evaluated the final student presentations of course projects. The rubrics were accumulated, and many students were interviewed for their self-evaluation. The data was analyzed by the assessment committee (two faculty members) and presented to the department for discussion at a department meeting.

Insights: What did we learn?

First, we learned that our students give good presentations. In general, they are confident speakers who enjoy talking about mathematics. They have a very good ability to generate very effective visuals.

Some general areas that students need to improve:

- Keeping presentations to the allocated time (in our case, 10–15 minutes)
- Including a good introduction to the talk
- Familiarizing the audience with notation and definitions
- Including a good conclusion to the talk.

As a result, we will continue to require students to include introductory and concluding comments in all their presentations, and we will encourage them to rehearse with a clock to ensure the time limits are met. We recommend that we continue to include oral presentations by our majors in as many of their courses as is purposeful.

Second, with regard to the assessment process, we learned that we needed a better rubric (e.g., to include the time requirement; see Appendix B). Also, it would be better to schedule the assessment earlier in the semester so that more effective feedback can be obtained from the students and given to the students. Additionally, students would not be burdened with so many competing end-of-semester requirements. The end of the semester also made it difficult for faculty to come together to discuss the student presentations.

Third, we gained useful insight into the assessment process as a result of this initial effort. One significant lesson learned in the process included the need to have a rubric briefing for all faculty participating in the assessment prior to the actual evaluations. Faculty raters agreed qualitatively about each presentation, but interpretation of what was meant by a numerical score varied among some raters more than is desirable. It is important to get consensus by the graders about what distinguishes a score of 1 from a score of 2. Additionally, there is no need that the one rubric be the department standard for every course. For our students' sake, there should not be wide variation from course to course with regard to format and standards for presentations, but instructors should be granted the flexibility to modify the rubric appropriately, based on their own emphasis. Videos were made for some of the student presentations, and those were marginally helpful to the students involved, but did not contribute significantly to the assessment results. Most importantly, getting all of our faculty together to talk about this issue helped create a common sense of purpose toward improving an aspect of our program that we feel is very important.

Appendix A. Initial Rubric

Oral Presentation Checklist

Student:_____

	Points	Comments
1. Introduction		
a. Introduces group members	/1	
b. Provides topic overview	/2	
c. States major result clearly	/2	
2. Presentation		
a. Presents correct content	/5	
b. Communicates with mathematical reasoning	/5	
c. Presents support for conclusions	/5	
3. Conclusion		
a. Reviews significant results	/3	
4. Style		
a. Quality of visuals	/2	
b. Apparent preparation (rehearsal)	/3	
c. Clarity of communication	/2	

General Comments: **Grade:** _____ **out of 30**

Appendix B. Revised Rubric

Oral Presentation Checklist

Name(s) _____

_____

	Points		
1. Introduction			
a. Introduce self (and teammates)	0	1	
b. Provide topic overview	0	1	2
c. State major results	0	1	2
2. Presentation			
a. Present correct assigned content	0	2	4
b. Communicate with correct mathematical reasoning	0	2	4
c. Present adequate support for conclusions	0	1	2
3. Conclusion			
Review significant results	0	1	2
4. Organization and Style			
a. Timing	0	2	
b. Quality of visuals	0	1	2
c. Clarity of communication, eye contact	0	1	2
d. Apparent preparation	0	1	2
Bonus: Creativity, appropriate humor	0	1	2

General comments: **Presentation Grade** _____

Keeping Assessment Simple

Jesús Jiménez and Maria Zack
Mathematics, Information Systems,
and Computer Science
Point Loma Nazarene University
San Diego, CA
jjimenez@ptloma.edu, mzack@ptloma.edu

Abstract. This case study describes an assessment plan created as part of an accreditation and five year departmental review process. It describes a simple but functional system for assessing major programs in Mathematics, Information Systems and Computer Science.

What did we hope to accomplish?

Point Loma Nazarene University (PLNU) is accountable to the Western Association of Schools and Colleges (WASC) as its accrediting agency. In the mid-90's PLNU went through a three-year cycle of document preparation and an accreditation visit. After that visit, the WASC visiting team informed the university that on its next visit it would put greater emphasis on planning and assessment. As the result of this information, PLNU has been attempting to develop a variety of planning and assessment tools to use in the ongoing activity of the institution. The Department of Mathematics and Computer Science made a first attempt at assessment in 1999 by conducting a five year department review. There were pluses and minuses to the process, but the main problem is that it did not develop a system of ongoing assessment and strategic planning.

In 2003 the department once again undertook a department review.[1] The team of nine faculty members in the department does not contain any experts in the area of assessment, and there is no one in the department who desires to make assessment part of his or her on-going scholarly work. With this in mind, we needed to design a system that would meet the expectations of WASC and our needs for planning and reviewing departmental effectiveness without becoming too cumbersome or time consuming. The review involved more than 100 pages of writing and documentation, but it also provided a reasonable template for future reviews (we should be able to update data and modify the text without rewriting the entire document). We believe that our department has developed a workable assessment and planning system that can carry us into the future.

What did we do?

PLNU is a small, liberal arts institution of approximately 2500 undergraduates. The Mathematic and Computer Science Department has roughly 100 students spread across three majors (Computer Science, Information Systems and Mathematics). The students in the department share some classes, and the faculty intentionally works to create a sense of unity among the 100 majors. To begin the assessment process, the Mathematics and Computer Science Department faculty sat down as a team and, over a period of weeks, developed a department mission statement and goals that represented the full department. This statement is shown in Figure 1 and is posted on our department website.[2]

---

[1] A full copy of the department review document can be obtained by emailing Maria Zack.
[2] mics.ptloma.edu/Department%20Assessment/Mission%20Statement%20and%20Goals.htm

Department Mission Statement: The Mathematics and Computer Science Department at Point Loma Nazarene University is committed to maintaining a curriculum that provides its students with the tools to be productive, the passion to continue learning, and Christian perspectives to provide a basis for making sound value judgments.

Department Goals: The goals of the Department of Mathematics and Computer Science are:

1. To prepare students for:
   - careers that use mathematics, computer science and management information systems in business, industry or government.
   - graduate study in fields related to mathematics, computer science and management information systems.
   - teaching mathematics and computer science at the secondary level.
2. To prepare students to apply their knowledge and utilize appropriate technology to solve problems.
3. To educate students to speak and write about their work with precision, clarity and organization.
4. To help students gain an understanding of, and appreciation for, the historical development, contemporary progress, and societal role of mathematics, information systems and computer science.
5. To integrate the study of mathematics, information systems and computer science with the Christian liberal arts.
6. To provide appropriate mathematical, information systems and computer educational support for any major area of study in this university.

Figure 1. Department Mission and Goals

Using these goals as a starting point, the department worked to develop an assessment plan. Our fundamental value was to develop an assessment program that was simple enough that it could be sustained over a number of years and was broad enough that the same assessment measures could be applied to all three majors. We looked at [1] to gather ideas about the success and failures of assessment programs and projects at other institutions. We found this resource invaluable in helping us find a place to begin.

The assessment plan that we developed for the department had to meet several criteria given to us by the university (including some externally verifiable measures). The institution required us to submit drafts of our plan in various stages of development. Having a deadline provided a good incentive for our team to wrap up discussion and make decisions. The assessment plan that we arrived at includes:

- An Alumni survey to be given once every five years. We used this tool in 1999 and modified it in 2003 to focus on some new initiatives in the department. This survey gathers information about the alumni and also asks attitudinal questions. Both the survey instrument[3] and results[4] are posted on our website.

- The creation of a Senior Seminar where all three majors gather together. In this seminar students hear talks by faculty and alumni and take the ETS Major Field Test in Mathematics (for Mathematics majors) or Computer Science (for Computer Science and Information Systems majors). The students also are required to give a 15 minute talk and write a short paper about a topic selected by the student in consultation with a faculty advisor. These talks and papers emphasize students' speaking and writing abilities and are graded by all department faculty members with the students being given the rubrics in advance.

- A comparison of our curriculum to national and state standards (these include MAA, ACM, ABET and AIS as well as the State of California standards for preparation of secondary school teachers).

- The use of external reviewers. We asked two faculty members in institutions similar to ours to take a look at our department review and all of the related data and to evaluate the curricular changes that we had made based on our assessment process. This was especially important in the area of Information Systems because we completely redesigned the curriculum. The two reviewers contributed valuable insights and suggestions that helped us to fine-tune our programs.

Details of the Assessment Plan are available in Appendix A and on our department web site,[5] as are the results of the ETS exams.[6]

After the preliminary round of assessment, we determined that we needed to make some changes to our assessment program.

- The Alumni Survey needs to be modified before we use it again. We gained some useful information from the data, but our assessment plan called for us to evaluate how the students rated PLNU's preparation of them for their next professional step (either a job or graduate school). Because of the small sample size, the standard deviations were large. So though the alumni told us that they were pleased with their preparation, we do not want to give too much weight to this conclusion.

- We had to wait on the university to get permission to add the Senior Seminar. In the interim, we gave the students the ETS test in a setting where it did not count as part of any course grade. The Math and CS students were given

---

[3] mics.ptloma.edu/Department%20Assessment/Alumni%20Survey.doc

[4] mics.ptloma.edu/Department%20Assessment/Alumni%20Survey%20Data.xls

[5] mics.ptloma.edu/Department%20Assessment/Assessment%20Plan.htm

[6] mics.ptloma.edu/Department%20Assessment/ETS_Exams.htm

the test in a class and the IS students took the test during an individually scheduled time with the department. We feel that most of the IS students did not take the test seriously. Since the students raw score will now be worth 20% of the grade for the Senior Seminar, this problem should be resolved. The Senior Seminar is graded on a pass/no pass basis.

What did we learn?

First and foremost, we learned the lesson of "keep it simple" when it comes to assessment plans. Our colleagues in other departments have still not begun to collect data because they created very elaborate and complex assessment plans and are overwhelmed by the amount of work needed to execute their beautifully crafted process. We are in the second year of data gathering and have already begun to learn useful information from our data.

Second, it is important to get full department "buy in" for the assessment plan. No one in our department sees this as her primary area of scholarship, yet because we built universally applicable goals and a department wide plan, all are willing to do their share of the work to get the assessment tasks accomplished. Certainly, the group approach to building an assessment plan is initially more time consuming but the long term benefits far outweigh the initial costs.

Third, our initial findings from the first full round of implementation of this very simple assessment plan were instructive. The general feeling in the department was that the data that we gathered resonated with faculty intuition about the status of the curriculum. Having the data to back up our intuition was very helpful in seeking administration level support to make needed changes. The key findings are:

- The students who graduated in 2000 or beyond feel better prepared in the areas of speaking, writing and the use of technology than those who graduated before 2000. Improvement of these skills was one of the goals of our 1999 curriculum changes. Though the Alumni Survey had some flaws, this information was very clear cut.
- The ETS MFT exams pointed out that our Mathematics and Computer Science majors were somewhat weak in a few specific areas (we found the ETS area specific sub scores much more useful than the overall scores). The findings of the ETS MFT for our Information Systems majors were a bit harder to use because the major has been a mixture of Computer Science and Business. There currently is no ETS MFT for IS so we had to test them using the CS exam. Overall, the MFT showed that our majors are performing well as compared to national norms and that greatly pleased the department's faculty.

- We needed to make only a few minor curricular changes to meet the current national standards for Mathematics and Computer Science (CUPM *Curriculum Guide 2004* and ACM, 2001). In 1999, during the department's last major "review" our curriculum for Mathematics and Computer Science was completely redesigned. Fortunately, the changes that were made at that time are consistent with the bulk of the recommendations in the current professional standards. However, our CS curriculum needed a few minor additions related to web technology. The national standards provided by ACM and ABET were helpful in crafting a new and more technical Information Systems major. Because we are a liberal arts institution, our restrictions on the number of non-GE units a student can take will prevent us from ever being able to apply for ABET accreditation, however the standards were helpful guidelines.

These findings were used in a variety of ways as we modified our curriculum.

- The Alumni Survey confirmed the need to continue our work with students in the areas of speaking, writing and the use of technology. We emphasize this across the curriculum with a capstone experience in the senior seminar. As much as the students do not like it when we are pushing them to speak publicly in their freshmen and sophomore years, it is an important part of their preparation for the professional world.
- We made some minor curricular adjustments in Mathematics. As indicated by ETS scores, we now require the students to take more applied mathematics. In the old curriculum, students were required to take four units of applied mathematics by choosing between Applied Mathematics (four units) and Mathematical Statistics (four units). In the new curriculum, students are required to take a total of ten units of applied mathematics by choosing several courses from Advanced Linear Algebra (two units), Numerical and Symbolic Computation (two units), Complex Analysis (two units), Applied Mathematics (four units), Discrete Mathematics (four units) and Mathematical Statistics (four units). In addition, we require all mathematics majors to take History of Mathematics (a suggestion made by a reviewer to better align our curriculum with our goals).
- We made some minor curricular adjustments in Computer Science. We now require all students to take Computer Architecture and have increased the number of units of Discrete Mathematics from two to four. Both of these changes were indicated by the ETS scores. We have also increased the number of units in the database course from two to three and have added a class in web

applications programming (recommended by outside reviewers and national standards).

- Our Information Systems curriculum was completely redesigned. With the addition of our new hardware lab, the major has become more "hands-on" and technical. Our outside reviewers agreed that this is consistent with the trend in Information Systems education and we hope that this will revitalize our major.

Next steps and recommendations

We are now in the second year of this assessment cycle; however this is the first year for the Senior Seminar because we had to go through the institution's academic course review process before we could add the course to our curriculum. We are currently fine-tuning the rubrics for the speaking and writing portion of the course and hope that we can craft an excellent course over the next few years. The initial response of the faculty and the students to the course has been positive.

Over the next couple of years, we need to modify the Alumni Survey and develop a more careful plan for follow up to increase the response rate. We will not be sending the survey again until 2008, so there is time for this work.

Our next significant assessment project is to develop a program for assessing our department's portion of the university's General Education program. Though our department has goals for our general education course, the university does not have clear goals for the overall general education program. This has made assessment difficult. PLNU students may take either one semester of Calculus or a course called Problem Solving to satisfy their general education requirement. We are currently giving an attitudinal survey as part of our assessment but need to see if we can develop tools for assessing problem solving skills. A fairly high level of department energy is currently going into this project.

Our department has learned some very valuable lessons in developing our assessment plan, the most essential ones are:

- Build your assessment from a clear set of goals that everyone in your department accepts.
- Make the assessment as easy as possible. Pick things that can be done in a straightforward manner and with a minimum of labor. Assessment can not be sustained if your faculty finds it too invasive or time consuming.
- Pick assessment tools that will back up your intuition. The faculty has a reasonably good sense of what is happening with students, but data can be very helpful in obtaining institutional resources to make the necessary changes.

Acknowledgements. We wish to thank the MAA, NSF and SAUM organizers and participants for valuable insights and assistance in developing our assessment program.

References

1. Gold, Bonnie, et al. *Assessment Practices in Undergraduate Mathematics.* Washington, DC: Mathematical Association of America, 1999.

Appendix. Learning Outcomes Assessment

Computer Science Major

Outcome #1 (Teach): *Graduates will have a coherent and broad-based knowledge of the discipline of computing.*

Means of assessment: Require students to take the ETS Major Field Test in Computer Science as the mid-term exam for the capstone course, Computer Science 481, Senior Seminar in Computer Science.

Criteria of success: 50% of our students achieve above the 25th percentile on the exam.

Outcome #2 (Shape): *Students will be prepared to give an oral technical presentation and a written summary of a topic in their field.*

Means of Assessment: Each student will be required to give a 20-minute oral presentation and a four page written summary of a topic in their field as a part of their participation in the Senior Seminar in Computer Science. The audience for this talk will include department faculty, fellow students and possibly some alumni. The students will be given the evaluation criteria in advance of their presentation and will be rated by the faculty using a rubric with a scale of 1 (outstanding) to 3 (unsatisfactory) in the following areas:

- Overall Content:
  - Technical information
  - Depth of information
  - Command of background material
- Oral Presentation:
  - Organization
  - Use of presentation tools
  - Notation
  - Exposition
  - Ability to field questions fro m the audience
- Written Summary:
  - Organization
  - Grammar and spelling
  - Notation
  - Clarity of writing
  - Bibliography and other supporting documentation

Criteria of Success: 80% of the students should have an average score of at least 2 in each of the major areas.

Outcome #3 (Send): *Computer Science graduates will be adequately prepared for entry into graduate school or jobs in the computing profession.*

Means of assessment: Alumni will be surveyed every five years. They will be asked at least the following questions:

If you have a job in Computer Science: On a scale of 1 to 5, 1 being outstanding and 5 being poor, how well do you think that the undergraduate Computer Science curriculum at PLNU prepared you for your work in the field?

If you are going to graduate school or went to graduate school: On a scale of 1 to 5, 1 being outstanding and 5 being poor, how well do you think that the undergraduate Computer Science curriculum at PLNU prepared you for graduate school?

Criteria of success: An average response of 2 for each question.

Mathematics Major

Outcome #1 (Teach): *Graduates will have a coherent and broad-based knowledge of the discipline of Mathematics.*

Means of assessment : Require students to take the ETS Major Field Test in Mathematics as the mid-term exam for the capstone course, Mathematics 481, Senior Seminar in Mathematics.

Criteria of success: 50% of our students achieve above the 25th percentile on the exam.

Outcome #2 (Shape): *Students will be prepared to give an oral technical presentation and a written summary of a topic in their field.*

Means of assessment : Each student will be required to give a 20-minute oral presentation and a four page written summary of a topic in their field as a part of their participation in the Senior Seminar in Mathematics. The audience for this talk will include department faculty, fellow students and possibly some alumni. The students will be given the evaluation criteria in advance of their presentation and will be rated by the faculty using a rubric with a scale of 1 (outstanding) to 3 (unsatisfactory) in the following areas:

- Overall Content:
  - Technical information
  - Depth of information
  - Command of background material
- Oral Presentation:
  - Organization
  - Use of presentation tools
  - Notation
  - Exposition
  - Ability to field questions fro m the audience
- Written Summary:
  - Organization
  - Grammar and spelling
  - Notation
  - Clarity of writing
  - Bibliography and other supporting documentation

Criteria of success: 80% of the students should have an average score of at least 2 in each of the major areas.

Outcome #3 (Send): *Mathematics graduates will be adequately prepared for graduate study, teaching and careers using Mathematics.*

Means of assessment : Alumni will be surveyed every five years. They will be asked at least the following questions:

If you have a job in industry: On a scale of 1 to 5, 1 being outstanding and 5 being poor, how well do you think that the undergraduate Mathematics curriculum at PLNU prepared you for your work in the field?

If you are going to graduate school or went to graduate school: On a scale of 1 to 5, 1 being outstanding and 5 being poor, how well do you think that the undergraduate Mathematics curriculum at PLNU prepared you for graduate school?

If you are in a teaching credential program or working as a teacher: On a scale of 1 to 5, 1 being outstanding and 5 being poor, how well do you think that the undergraduate Mathematics curriculum at PLNU prepared you for teaching?

Criteria of success: An average response of 2 for each question.

Information Systems Major

Outcome #1 (Teach): *Graduates will have a coherent and broad-based knowledge of the discipline of Information Systems.*

Means of assessment : Require students to take the ETS Major Field Test in Computer Science as the mid-term exam in IS 481, Senior Seminar in Information Systems.

Criteria of success: 50% of our students achieve above the 25th percentile on the exam.

Outcome #2 (Shape): *Students will be prepared to give a written summary of a topic in their field.*

Means of assessment : Each student will be required to give a 20-minute oral presentation and a four page written summary of a topic in their field as a part of their participation in the Senior Seminar in Information Systems. The audience for this talk will include department faculty, fellow students and possibly some alumni. The students will be given the evaluation criteria in advance of their presentation and will be rated by the faculty using a rubric with a scale of 1 (outstanding) to 3 (unsatisfactory) in the following areas:

- Overall Content:
  - Technical information
  - Depth of information
  - Command of background material
- Oral Presentation:
  - Organization
  - Use of presentation tools
  - Notation
  - Exposition
  - Ability to field questions fro m the audience
- Written Summary:
  - Organization
  - Grammar and spelling
  - Notation
  - Clarity of writing
  - Bibliography and other supporting documentation

Criteria of success: 80% of the students should have an average score of at least 2 in each of the two main areas.

Outcome #3 (Send): *Management Information Systems graduates will be adequately prepared for entry into the information systems profession.*

Means of assessment : Alumni will be surveyed every five years. They will be asked at least the following question:
1. If you have a job in computer science: On a scale of 1 to 5, 1 being outstanding and 5 being poor, how well do you think that the undergraduate Management Information Systems curriculum at PLNU prepared you for your work in the field?

Criteria of success: An average response of 2.

Surveying Majors in Developing a Capstone Course

M. Paul Latiolais, Joyce O'Halloran,
Amy Cakebread

Department of Mathematical Sciences
Portland State University
Portland, OR
latiolaisp@pdx.edu, joyce@pdx.edu

Abstract. The Department of Mathematics and Statistics at Portland State University surveyed its undergraduate majors (and some graduate students) to gather information for the development of a senior mathematics capstone course. Students were asked about the importance of a list of performance objectives and their perceived competency in each objective. Demographic information was also collected. Results and analysis of the survey are presented, as well the role of these results in departmental decisions.

What did we hope to accomplish?

The design and implementation of courses in the Department of Mathematics and Statistics at Portland State University has traditionally been based on what mathematics departments in research universities do and what faculty in the department think might be needed. Removal of courses from the curriculum has been similarly based on what faculty believe is needed. Asking the students what they need in a controlled manner has not been done.

After lengthy discussions over several years, the department agreed in June of 2001 on a list of student learning objectives for the major. Those learning objectives were in six categories: *Mathematical tools, Connections, Technology, Communications, Independent learning,* and *Attitudes*. In mapping these objectives to the curriculum, the department discovered that many of the objectives were not effectively addressed by the curriculum in place at that time. The learning objectives that were not being met (see Figure 1) were largely in three areas: *Connections, Communication,* and *Independent learning*.

At the same time, the dean of the College of Liberal Arts and Sciences urged the department to develop an assessment of the major as part of the university's assessment initiative. While developing strategies to assess student learning objectives, the department realized it needed new strate-

---

Connections:
    Applications: Awareness of applicability of math in other
       disciplines
    History: Familiarity with historical/social contexts of math-
       ematics
    Contexts: Ability to make connections in math from one
       context to another
    Models: Ability to build and use mathematical models of
       concrete situations or real phenomena
    Statistics: Ability to use data and statistical techniques to
       solve a problem or make a supportable conclusion

Communication:
    Delivery: Proficiency in oral and written communication of
       mathematics to peers as well as to people with less math
       background
    Teamwork: Ability to work as part of a team to do math

Independent learning:
    Independence: Proficiency as an independent and critical
       thinker
    Library: Ability to use the library and other non-classroom
       resources to solve a problem in math
    Questioning: Ability to ask the right questions to learn
       something new or apply something known to a new sit-
       uation

Figure 1. Unmet Student Learning Objectives

gies to assess students at the senior level. The department has developed strategies to assess student skills as they begin abstract math courses, but the department does not have a way to measure the value added between their first abstract math course and completion of the undergraduate program. We realized that we could develop an end of program assessment and address student learning objectives not yet covered with a senior capstone experience.

In order to begin the design of a senior capstone course, we undertook a systematic survey of current and past undergraduate mathematics majors' needs. The intent was to ask students what they thought was most needed in their studies of mathematics and whether they would voluntarily take such a capstone course addressing those needs.

With the help of a graduating senior mathematics major, we designed, piloted and administered a survey of math students. The survey focused on ten particular learning objectives from the departmental list that the department felt were not being well addressed or well assessed (see Figure 1). The survey asked students how proficient they felt they were in the identified learning objectives and how important they thought each objective was. The survey also described a potential model for the capstone course and asked students for feedback on the course design and whether they would take such a course.

A pilot survey was administered in the winter of 2002. It provided valuable feedback on how students perceived the questions that were asked and how to improve the survey to get information we needed. The survey was subsequently redesigned, changing some of the column formatting and the phrasing of some questions. The survey was administered at the beginning of fall term 2002. The results of the survey will be discussed in subsequent sections.

Survey Responses and Analysis

The survey was administered in two classes: a junior level advanced calculus class and a senior real analysis course. The advanced calculus course is required of all majors, so we thought we could get a good representation by surveying that class. The vast majority of students in the real analysis course are typically seniors, so we could be sure to get their voice by surveying that class. As advanced calculus is a prerequisite for real analysis, we would not get any overlap.

Student Demographics

A total of 40 students were surveyed; 30 undergraduate and 10 graduates. As Figure 2 shows, 12 students identified themselves as female; 27 identified themselves as male. One student did not declare a gender. For the purposes of

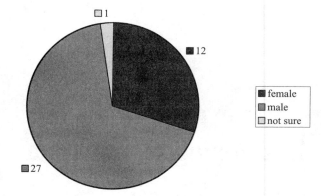

Figure 2. Gender Distribution

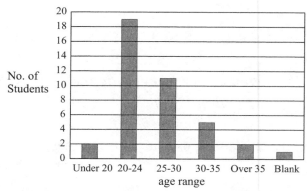

Figure 3. Age Distribution

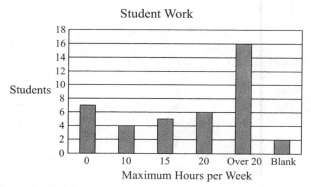

Figure 4. Work hours

statistical analysis we decided to declare that student as "not sure." As can be seen in the Figure 3, the majority of students were between the ages of 20 and 24. The work chart (Figure 4) shows that the majority of students worked over 15 hours per week At least one student worked full time.

Student Responses

Students were asked about the importance of student learning objectives in three areas: *Connections, Communication* and *Independent learning* (see Figure 5). The most interesting result of the survey was that the majority of students felt

Response	Applications	History	Connections	Models	Statistics	Delivery	Teamwork	Independence	Library	Questioning
Blank	0	0	3	1	1	1	2	1	3	1
Skill not needed	1	0	0	0	0	0	1	0	1	0
Very little of the skill is useful	2	7	0	1	3	0	2	0	3	0
Some aspects of this skill are useful	9	14	4	9	5	3	7	1	9	2
Much, though not all, of the skill is useful	13	**15**	12	14	14	12	10	8	10	13
Essential skill	**15**	4	**21**	**15**	**17**	**24**	**18**	**30**	**14**	**24**

Figure 5. Student responses to question of "importance" of ten learning objectives.

Response	Applications	History	Connections	Models	Statistics	Delivery	Teamwork	Independence	Library	Questioning
Blank	0	1	3	1	1	1	2	1	3	1
Do not have this ability	0	6	0	2	1	0	1	0	2	1
Some understanding	9	7	3	5	6	4	4	1	5	0
Mostly OK	8	**16**	10	11	**14**	10	7	9	**13**	15
Proficient for my needs	**19**	8	**19**	**18**	**14**	**24**	**17**	**20**	11	**18**
Expert	4	2	5	3	4	1	9	9	6	5

Figure 6. Student responses to question of "skills" of ten learning objectives.

each student learning objective was important. (The tables in Figures 5 and 6 are labeled by key words of the learning objectives. Highlighted cells signal the maximum number of responses. The master key is supplied in Figure 1.)

The most important objective to the students was "Proficiency as an independent and critical thinker." The second most important objective was "Proficiency in oral and written communication of mathematics." The lowest scoring objective was "Familiarity with historical/social contexts of mathematics." It is not surprising that students valued this objective less. What is surprising is that they still thought it was important. The department's mathematics history course is taught only in the summer and only by non-regular faculty. The course is not required of majors. The social context of mathematics is not well addressed either. Students infrequently encounter the relevance of the historical or social contexts of mathematics.

As Figure 6 shows, students felt that they were most skilled in "Proficiency as an independent and critical thinker", the same objective they felt was most important. Students rated "Ability to work as part of a team to do math" second in their perceived skill level. A focus group study of two years ago told us why students might feel they are good at working in groups; the department's atrium is full of students all day long collaborating on mathematics together. The atrium may be our greatest asset. The objective scoring lowest in skill level was "Familiarity with historical/social contexts of mathematics." This result is not surprising from our earlier comments. The department does not offer much in these areas, hence students do not have the skill.

By subtracting the average responses on skill from those on importance, we get a sense of areas where students feel the need for the most improvement. The difference is the

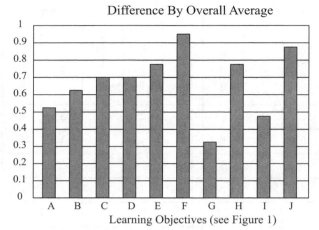

Figure 7. Differences between importance and skills.

highest for "Proficiency in oral and written communication…" and the lowest at "Ability to work as part of a team to do math." We conclude that students feel that their oral and written communication need the most improvement. Other skill/importance differences can be seen in the chart below. (Notice that even where students felt they were skilled in a particular objective, they still expressed a need for further improvement.)

Course Comparison

While both the junior level advanced calculus and senior level analysis students felt "Proficiency as an independent and critical thinker" was the most important skill, the students surveyed from advanced calculus did not rate the skills overall as important as the analysis students did. When comparing their skill level, the students from advanced calculus felt more confident about their skills than the analysis students. (Makes you wonder what happens to students to make them less confident about their skills between advanced calculus and analysis classes!) Overwhelmingly, the analysis students felt they lacked "Proficiency in oral and written communication."

Gender Differences

Females rated "Awareness of applicability of math in other disciplines", "Familiarity with historical/social contexts of mathematics", and "Ability to build and use mathematical models" as the areas where they would like to improve on most. Males, on the other hand, rated these as the lowest differences. That is, the men felt that the rest of the areas needed more improvement. Females felt they had sufficient skills in "Ability to work as part of a team " and "Ability to use the library and other non-classroom resources to solve a problem in math." These gender differences may be attributable to the women being older than the men.

Graduate versus undergraduate

The graduate students felt "Proficiency in oral and written communication of mathematics to peers as well as to people with less math background" and "Proficiency as an independent and critical thinker" were most important, while the undergraduates felt "Proficiency as an independent and critical thinker" and "Ability to ask the right questions to learn something new or apply something known to a new situation" were most important. The graduates felt "Proficiency in oral and written communication" was where they needed improvement. The undergraduates felt "Ability to ask the right questions" was most needed. Neither the graduates nor the undergraduates felt they had as much skill as they would like in any category.

Open Ended Questions

After asking students to rate importance and skill level on the objectives, the survey asked them specifically for areas in which they would like to improve their proficiency, from the list and in other areas not on the list. We also asked them why it was important to become more proficient in these areas. Of the 40 respondents, 27 students answered this question. Of objectives from the list, "Proficiency in oral and written communication" was the favorite with 13 out of 27. When asked about areas not on the list, only six students out of 40 answered. Writing proofs was the favorite choice here with four of them. These answers seemed to reflect a desire of students to see more of the objectives we listed in the survey as part of the departmental curriculum.

Lastly, the survey asked students to critique a senior capstone course described to them as follows:

"We are proposing a two-term course—worth two credits per term. Topics for the two terms would include the history of mathematics, the application of mathematics, and the applicability of mathematics to specific disciplines. Each student would be required to give oral presentations and written reports. All students would help give feedback to presenters throughout the entire process.

During the first term, students would work in groups to do a variety of short research activities. The second term would consist primarily of two activities:

1 Assisting students going through their first term

2. Creating an independent research project around one of the topics addressed in the previous term (or a topic suggested by the faculty).

Second term students would present their work at a public event. This two-term course could potentially satisfy the University Studies capstone, but approval would be discretionary and would require the addition of a community service component."

They were also asked if they would take such a course. Only 11 students responded to those questions. Critiques were not significant, although many felt that the course needed to offer more course credit and should satisfy the university's capstone requirement. Nine of the 11 students said that they would take such a course.

Survey Issues

The ethnicity category was omitted from the analysis because of the lack of diversity in those surveyed. (75% of the students surveyed were white.) Since only three surveys were from non-math majors, this category was also omitted. Due to the invalid and often incorrect answers to the question of the number of credits a student had, this category was omitted. GPA estimates were mostly above a 3.0 and also were omitted from the analysis due to the lack of diversity.

What did we learn?

In any study like this, results include the questions that arise. One immediate question we had was, "If we gave the same survey to linear algebra students, would the responses be different?" The two courses surveyed are very difficult for our students. Their perceived need for these skills may be underlined by the difficulty of the material they are trying to learn.

The Senior Capstone

Our declared purpose of the survey was to help us design the senior capstone course. Based on the survey results, what should be included in the senior capstone course? Keeping in mind that students considered all of the listed objectives important, they all need to be incorporated into this course to some degree. The survey indicated that students feel confident about their communication skill level. At the same time, their responses also indicated that they would like to be more proficient in this area. Hence written and oral communication will drive the format of this course, with students presenting foundational information as well as solutions to problems. With "proficiency as an independent and critical thinker" rated the highest in importance, a discovery format would best serve the students. The discovery process could include meta-cognition addressing the objective of "asking the right questions to learn something new." As far as course content, it will be important to explore connections within mathematics and between mathematics and other disciplines.

Because several student responses emphasized the desire for a senior capstone course which would satisfy the university's capstone requirement, we are incorporating the required community-based component. Specifically, the course (offered for the first time as a 2-quarter sequence starting January 2004) will include presentations to inner city high school students. Integrating community-based learning into this course will involve readings and discussions about the role of mathematics in our society, thereby addressing the "social context of mathematics" learning objective.

A one-quarter "pilot" capstone course (with no community-based component) was offered Spring 2003, in which students explored applications of mathematics independently, made presentations in class, and wrote a final paper. Experience with the pilot course demonstrated the need for a 2-quarter experience, both for exploration and in order to incorporate community-based learning.

Assessing Written and Oral Communication of Senior Projects

Kevin Dennis

Department of Mathematics and Statistics
Saint Mary's University of Minnesota
Winona, MN
kdennis@smumn.edu

Abstract. In an evolving process to assess its majors, the mathematics department decided to use Senior Project presentations and reports as instruments to evaluate the department's goal of developing the skill of mathematical communication. The department developed rubrics for oral presentations and written reports that were piloted on a recent class of graduates. The findings and recommendations from these rubrics are discussed.

Introduction

Saint Mary's University of Minnesota is a private Catholic university conducted by the De La Salle Christian Brothers with campuses throughout the world. The Winona campus offers a coeducational, residential, liberal arts program with an undergraduate student enrollment of about 1,300 and graduate programs in a variety of disciplines. The Department of Mathematics and Statistics at the Winona campus, with six full-time faculty members, offers a major and minor in mathematics and a minor in statistics. The department graduates between four and seven mathematics majors per year.

The mathematics department has a comprehensive assessment plan for its majors. We use Junior and Senior Assessment exams to assess students' content knowledge and Senior Exit Interviews and a Nature of Mathematics Survey to assess students' attitudes towards mathematics. In addition, to assess students' mathematical communication skills, Senior Projects and collection of homework from proof writing courses are evaluated. In this paper, we will concentrate on the assessing of written and oral communication of the mathematics major's Senior Projects.

All members of the Saint Mary's Mathematics department agree that the communication of mathematics is one of the most important skills a major should gain as an undergraduate. Hence, the department has made the oral and written communication of mathematics a learning outcome for its majors. To assess this learner outcome, the department has devised an assessment tool to evaluate the effectiveness of its program to this outcome.

Background

At Saint Mary's University of Minnesota, every senior mathematics major must take a course entitled "Senior Seminar." This one-credit course is offered during the spring semester. The focus of this course is an independent research project called the Senior Project, which counts for 70% of the course grade. The other 30% of the grade is determined by participation in class discussions and the Senior Assessment Exam. To help prepare students for the Senior Project, the instructor of the course selects mathematical journal articles for students to read and discuss. These articles aid students in learning about the format of a mathematical paper. After reading several articles, students then choose a topic of interest to research. Working with members of the department, students select topics that are of interest to them. Students typically choose topics involving mathematical modeling or mathematics education.

After learning about their topic, students must present their project in a 25-minute talk with a 10-minute question and answer period afterwards. All faculty and students of Saint Mary's are invited to this talk. Historically, the oral presentations were evaluated using questions found in Appendix A, but on a Likert scale of 1 (Low) to 7 (High). The department felt that the Likert scale was too subjective and that a more quantifiable scale should be employed. Hence, the scale found in Appendix A was created.

After presenting their topic, the students must also write a 15–20 page technical report on their project. Note that the term "technical report" is used instead of paper to denote the format of the paper specific to the discipline of mathematics. For past written reports, only the instructor of the Senior Seminar course read the reports of the Senior Projects, mainly to determine the student's course grade. Since the department decided to assess the students' written communication, the department created a rubric to assess the technical reports.

Method

To assess the oral presentations, any faculty member, regardless of discipline, attending the Senior Project presentations is given the form found in Appendix A. Typically, there are five to seven faculty members attending each presentation. After a student's question and answer period, the faculty members are given a few moments to fill out the form while the next student prepares for his or her talk. In addition to answering the seven questions, faculty members also write comments about the mathematics being presented, comments to the student and comments to the instructor. After the assessment forms are collected, the students receive the averages of their scores along with any appropriate comments about their mathematics and their presentation. The course grade for the oral presentation is determined by the instructor of the Senior Seminar course, who may or may not use this evaluation in determining the grade.

To assess the written technical reports, the department decided to have three faculty members read each report. For a particular paper, the three faculty members, with one being the instructor of the course, fill out the form found in Appendix B. The scores from these questions are tallied, usually after the semester is concluded, so the students do not receive them. The course grade for the written technical report is determined solely by the instructor of the Senior Seminar course. The department then meets at the end of the semester to discuss all assessment items, including the oral presentations and written reports.

Findings

The rubrics for oral presentations and written reports in the forms found in Appendices A and B were piloted during the 2002–03 school year. Due to this fact, some of the findings from the department are about the assessment forms themselves.

Student Communication Skills. With the oral presentations, the scores for the seven questions on average ranged between five and six for each student. With these scores, the department agreed that the students for the most part were good presenters. For example, they were organized using Power Point presentations and had very good diction. Unfortunately, the students did not give good mathematical presentations, often having trouble communicating their knowledge of the topic. The students had a tendency to give lengthy explanations for simple mathematical concepts, while quickly explaining the more difficult concepts needed for their projects. In addition, the students used notation in their talks without explaining the meaning. The department believes that the mathematical explanations distinguished the Senior Project presentation from any other presentation that a student may have done in college. Even so, this attitude did not carry over to the students since most students still treated the Senior Project presentation as just another presentation. For example, the students put more time into the appearance of the presentation than developing the mathematics required to demonstrate their knowledge of the topic.

With the written reports, the scores on average ranged between 3 and 4 for each student. With these scores, the department felt that the students met only the minimal expectations of the project. The students did not research the history of their topic, but instead just found the minimal amount of formulas and theorems needed to solve their problem. As with the oral presentations, the students had trouble with communicating the mathematics effectively. Many times, the students would show how to place numbers into a certain algorithm, instead of showing how the algorithm is generated.

Rubrics. The department found some problems with the rubrics themselves. For the oral presentation rubric, the first problem was that the standard made it difficult to judge high scores. Note that scores of 4 through 7 on the standard all dealt with just needing minor changes in the talk. The second problem was the interpretation of the rubric standard. Though the students' scores averaged between 5 and 6, some students had scores ranging from 3 to 7 for the same question. This raises the question of inter-rater reliability. During the discussion of these issues, some faculty members indicated that they thought the rubric was to be used in determining the course grade and, hence, scored higher than the standard called for. Another interpretation problem was that mathematically "weaker" topics were given lower scores, though the level of the topic is not incorporated into the rubric. The department's only trouble with the written report rubric was that it was hard to apply to mathematics education projects.

Recommendations

The department responded to the findings from the assessment by recommending changes to the format of the Senior Seminar course. To aid in some of the communication issues of the oral presentation, the department suggested that the Senior Seminar instructor incorporate a practice session in the class. The students would present their topic to the Senior Seminar class and be graded on that presentation, then give the talk to the Saint Mary's community. To aid in the minimalism that existed in the students' reports, the department suggested that the instructor encourage "stronger" mathematical topics, with even the possibility of restricting the choice of topics. In addition, the instructor should encourage the students to start their projects during the fall semester, instead of waiting until the spring.

One factor against the success of increasing the communication skills via changes in the Senior Seminar class is that the mathematics department encourages the rotation of its upper level courses to all department members. With this change, some continuity in the Senior Seminar class is lost, so information gained from one year's class may not be reflected in the next year's Senior Seminar class. On the other hand, an encouraging note is that the university has gone to a full year planning schedule. In the past, the uni-

versity scheduled its courses on a semester-to-semester base, which meant that the Senior Seminar instructor was not chosen until middle of the fall semester. Now the Senior Seminar instructor is known before the school year starts, so the instructor can work with the senior mathematics majors informally in the fall, preparing them for their Senior Project. This means that the Senior Seminar class in the spring can be used to discuss the communication of mathematics more than researching mathematics.

Besides suggesting changes in the Senior Seminar class, the department set some goals in order to elevate the students' mathematical communication skills. The main goal was to give students more opportunities to see examples of mathematical communication, both orally and in written form. To see more written examples, one recommendation was to have all instructors of upper-level mathematics classes have students read journal articles and write reports over these articles. For oral communication, the department set a goal to encourage more students to go to conferences and have more outside speakers present at Saint Mary's University of Minnesota. One idea discussed was to have the top two Senior Seminar presenters be awarded free trips to the local mathematical conferences held in April.

With the information obtained through this assessment process, the mathematics department at Saint Mary's University of Minnesota learned that while its majors present well and write well for general topics, the technical communication of mathematics is lacking. The department has set goals to have the students see and read more examples of good mathematical communication in and outside the classroom. Hopefully, through these examples, the students' mathematical communication skills will increase over time.

Appendix A. Sample Presentation Assessment Form

7 – Excellent – ready for presentation at a conference
6 – Very Good – needs little changes before presenting at a conference
5 – Good – needs a few minor changes before presenting at a conference
4 – Satisfactory – needs minor changes before presenting at a conference
3 – Poor – needs few major changes before presenting at a conference
2 – Very Poor – needs major changes before presenting at a conference
1 – Unsatisfactory – presentation has little or no value

Questions:

1. How well did the student explain the purpose of the project?

2. How well did the student organize the material?

3. How neat was the student's presentation (use of overheads, PowerPoint, etc.)?

4. How would you judge the student's presence (voice, delivery, etc.)?

5. How well did the student answer/respond to questions and comments?

6. How well did the student communicate his/her understanding of the mathematics?

7. Overall, how would you rate this presentation compared to other presentations you have seen (before today)?

Comments on the mathematics:

Comments to the student:

Comments to the instructor:

Appendix B. Sample Report Assessment Form

5 – Excellent – needs minor revisions in a few places (e.g., 2 or fewer changes)
4 – Good – needs minor changes throughout report (e.g., 1 per page)
3 – Satisfactory – needs a major change or many minor changes (e.g., over 2 per page)
2 – Poor – needs major revisions throughout report
1 – Unsatisfactory – has little to no value

Presentation of Report

1. How well written is the report (e.g., correct grammar, spelling, etc.)?

2. How well does the student present graphics/figures/equations (e.g., placement, neatness, etc.)?

Technical Report

3. How well does the student explain the purpose of the project?

4. How in depth does the student explain the history of the problem?

5. How well does the student explain the background mathematics needed to understand the problem?

6. How well does the student critique the model/study/topic?

Content

7. How in depth does the student explore the mathematics/statistics/research?

8. How accurate are the mathematical/statistical statements?

9. How well does the student justify the mathematical/statistical statements?

10. How consistent and effective is the student's use of notation?

11. How effectively does the student use examples to clarify points made in the paper?

Overall – Report

12. How well does the student demonstrate that he/she understands the mathematics/statistics/research?

13. How well does the paper compare to other reports seen before this years?

Overall – Topic

14. The level of the topic appears obtainable with someone in a background in:
    1 – Freshman level mathematics and statistics (Calc I/Calc II/Intro to Stats)
    2 – Sophomore level mathematics and statistics (Calc III/Linear/Probability)
    3 – Junior/Senior level mathematics and statistics (First course in specialized areas)
    4 – Graduate level mathematics and statistics (Second course in specialized areas)
    NA – Educational Research Project

15. The level of originality of the topic appears to be:
    1 – Using known techniques to solve a problem found in a lower level course
    2 – Using known techniques to solve a problem covered in an upper level course
    3 – Using known techniques to solve a problem partially covered in an upper level course
    4 – Using unknown techniques to solve a problem partially covered in an upper level course
    5 – Using techniques to solve a problem unknown to the student
    6 – Using techniques to solve a problem unknown to the mathematical community or contributing new literature to the educational community

Assessing the Mathematics Major: A Multifaceted Approach

Brian Hopkins, Eileen Poiani, Katherine Safford

Department of Mathematics
Saint Peter's College
Jersey City, NJ

bhopkins@spc.edu, poiani_e@spcvxa.spc.edu,
safford_k@spcvxa.spc.edu

Abstract. The Saint Peter's College Department of Mathematics has developed a multifaceted assessment plan to better understand and improve student learning in the mathematics major. Key elements include departmental objectives, student portfolios, analysis of transcripts, and surveys of alumni; the latter three will be explored in detail in this case study. Our goal is to make informed decisions about potential changes in the major and provide a solid foundation of student learning data going forward.

Background

Saint Peter's College (SPC), founded in 1872, is a Jesuit liberal arts college of approximately 3,700 students located in Jersey City, New Jersey. The SPC Department of Mathematics has a long history of attention to assessing student learning, evidenced by Eileen Poiani's analysis of the developmental program [1].

The current effort is focused on the major program. How successful is the current course sequence? How is student success measured: in-class? after graduation? We decided that a multifaceted approach was required, considering classroom activities, college records, and alumni.

This assessment of student learning focusing on the major was requested by the academic dean in 2001. Conveniently, this coincided with the formation of the first "class" sponsored by the Supporting Assessment in Undergraduate Mathematics (SAUM) program of the Mathematics Association of America. Eileen Poiani, Katherine Safford, and Brian Hopkins were the three SPC faculty who participated in the first assessment workshop (three meetings) in 2002 and 2003. The leadership and colleagues in the program contributed greatly to our developing assessment efforts.

Description

The department began its assessment endeavor in 2001 with a department meeting dedicated to discussing our mission statement and goals (See Figure 1). We developed objectives for student learning by considering ways to complete the sentence *"The student, upon completion of the major, will be able to ..."* without asking for specific content knowledge such as "apply the Sylow theorems." The result

---

Department Mission Statement: The Saint Peter's College Department of Mathematics seeks to develop in our students the level of mathematical competence appropriate for their educational goals, to foster appreciation of mathematics as part of human culture and in relation to other fields of study, and to encourage the intellectual growth of students and faculty.

Department Goals: The Department of Mathematics seeks to equip all Saint Peter's College students with quantitative reasoning and critical thinking skills that enable them to be informed participants of an increasingly technical society. For majors, the department seeks to inculcate a significant amount of mathematical content and maturity in several areas; to foster logical thinking, creative problem solving, and precise communication; and to encourage the application of this content and methodology both within and outside mathematics.

Figure 1. Mission and Goals

is displayed in Figure 2. Then we discussed methods to measure the success in meeting these objectives, both practices currently in place and future possibilities (Figure 3). Cross-indexing the objectives and methods against specific major courses constituted our report to the college, which was well received. Appendix A shows the relation of objectives to courses and to assessment methods.

In 2002, as participants in the SAUM workshop, we decided to expand the department assessment endeavor by developing additional projects, using student transcripts, alumni surveys, and portfolios. The last major change in coursework for the major occurred in 1978, so we used 1980 as a baseline for our efforts. Analysis of math major transcripts from 1980 to 2003 (a total of 90 records) allowed us to see changing patterns in our students. With which courses did students begin their mathematics study? What were the other academic pursuits of our majors? Did they follow the system of prerequisites? For those who did not, how was their grade point average affected?

The Saint Peter's College Department of Mathematics has a long and rich history of communication with alumni, primarily through an annual newsletter and written responses. In the spring of 2003, we coupled the normal communication and a survey requesting information about particular classes and "their value to your post-graduation employment and educational experiences." (The survey instrument is Appendix 2 of our Case Study report on the SAUM web-

---

1. In-class tests and quizzes
2. Take-home tests
3. Homework
4. Collaborative projects done in small groups
5. Computer solutions and simulations
6. Papers
7. Presentations to faculty outside of class (of computer work, projects, papers)
8. In-class presentations (of homework, computer work, projects, papers)
9. Poster Day and Pi Mu Epsilon student presentations
10. Graduate surveys
11. Discussions with faculty colleagues

Figure 3. Means of Assessment

site.[1]) Some 84 surveys were successfully delivered, and 24 were completed and returned.

In order to have evidence of longer-term student development, we have begun to collect work for student portfolios. Colleagues with portfolio experience strongly recommended against collecting too much paperwork, which can easily become unwieldy. We are collecting one document per student per major class, with specifics determined by the instructor. In discrete mathematics, for instance, I asked students to submit "a proof you're proud of." After starting with the 2002–2003 sophomore classes, and then expanding to all major courses, we will have portfolio documents from twelve classes by spring 2004.

Also, we have incorporated the student learning objectives in regular surveys of students and instructors for each class. In addition to the college's student evaluation, we have students rate how much each departmental objective was addressed in the course (making clear that there is no expectation that every class will address every objective). Likewise, the instructor fills out the same survey and provides additional narrative on how various objectives were addressed and what methods were used for assessment.

Insights

Beginning with a full faculty discussion helped involve all instructors and prevent our assessment efforts from becoming compartmentalized within the department. The discussion of methods of assessing student learning was very enlightening and gave several faculty new ideas that have since been incorporated. For example, at least one instructor started having students give in-class presentations on supplementary articles, and several began using Excel much more widely in various courses. Student and instructor sur-

---

1. Think logically and analytically
2. Demonstrate a strong level of mathematical maturity
3. Solve problems creatively
4. Apply technology in solving problems
5. Produce concise and rigorous mathematical proofs
6. Appreciate the history of mathematics as a human endeavor
7. Recognize the interconnection of various fields in mathematics
8. Construct mathematical models
9. Apply mathematical content to other disciplines
10. Transfer mathematical thinking (logic, analysis, creativity) beyond cognate fields
11. Access relevant resources when posing and answering mathematical questions
12. Read and assimilate technical material
13. Produce cogent mathematical exposition
14. Communicate technical material effectively at an appropriate level
15. Succeed in mathematics graduate study, K-12 mathematics instruction, or other careers requiring computational or analytic skills

Figure 2. Student Learning Objectives

---

[1] www.maa.org/saum/cases/StPeters_A.html

veys based on the objectives have kept the assessment program alive in our regular work of teaching.

The transcripts partitioned easily into four six-year periods (see Appendix B). Of particular interests are the trends into the most recent period, 1998–2003. Encouragingly, the number of mathematics majors is increasing. Also, the proportion of female majors is now slightly more than 50%. In terms of minors and double-majors, the only frequent combinations are with computer science and education. But while more than half of the mathematics majors in the 1980s also focused on computer science, the last period shows an equal percentage of students focusing on education as computer science. Looking at our current students, it appears that these changes will continue: fewer students studying mathematics and computer science (especially since the Department of Computer Science was recently moved to the School of Business), more students studying mathematics and education (with the intent of teaching K–12).

Transcripts also show a change in how students begin the major. Earlier, most students (about 75%) started with the calculus sequence dedicated to math and science majors or at a higher level. For the most recent period, that figure is down to about 60%, which means that some 40% of our majors of the last six years started with elementary calculus, finite mathematics, or math for humanities. This is great news in terms of recruiting majors, but makes completing the required coursework in four years challenging, as our upper division courses are offered every other year. It is not surprising, then, that there has also been a significant rise in the percentage of students who did not follow our prerequisite system. However, there was also no statistically significant difference in the grade point averages of students who did or did not follow the prerequisites, another provocative datum.

On a practical note, acquiring and analyzing the desired transcripts was tedious. Getting the proper records from the registrar took time, especially for students whose records were on hold because of outstanding financial issues. Due to FERPA regulations, student workers could not help enter or analyze the data. More frustrating, the format of the college's records did not allow us to track students who had left the major. More complete tables of the results mentioned above are provided in Appendix B.

The alumni surveys gave consistent feedback on which courses proved helpful or less so for the post-college careers of former math majors. Several courses were deemed "very valuable" by more than half of the respondents: mathematical modeling, probability, various calculus courses, statistics, linear algebra, and discrete mathematics. Some of these same classes were listed as "courses you think were most helpful in your career choice." On the other end of the scale,

only two classes were deemed "not valuable" by more than half the respondents: numerical analysis and modern algebra. Common responses for "courses you think were least helpful in your career choice" were modern algebra and advanced calculus (real analysis). More complete results are provided in Appendix C.

Surveys can also be frustrating, trying to balance the information you want with a form that is not overwhelming, waiting for the results, and dealing with low return rates. Although our 29% return rate is respectable, we are hesitant to place too much weight on the feedback of 24 alumni. Looking at a particular result, how do we respond to the negative feedback on modern algebra? Could the class be altered to more directly tie in to high school algebra? Should we expect it to play a role in the vocations of graduates, or does it serve a different function? Also, the course-by-course structure may work against a holistic view of the major; one respondent attached a letter explaining how the problem-solving skills and critical thinking developed by the major were very helpful, but she could not tie them to particular classes.[2] Nonetheless, we were very pleased to receive her letter.

Next Steps

There remains more to do on all of our assessment projects. The transcript data can be mined to address more questions. There are qualitative responses on the survey that should be compiled and summarized. Student portfolios are accreting at a steady rate, but we will have to devise a rubric before they can yield helpful information (the first portfolios will be completed spring 2005).

Our next large goal is using all of the assessment data to help make informed decisions about the major: courses, sequence, prerequisites, etc. Another resource guiding any potential change will be the *CUPM Curriculum Guide 2004* [2]. This reconsideration of the major program is planned for 2005.

Looking farther into the future, continuing portfolios, transcript analysis, attention to goals, and subsequent surveys will provide invaluable longitudinal data. An ongoing attention to assessment will give the department an even richer foundation for improving student learning in the mathematics major at Saint Peter's College.

Acknowledgements: Katherine Safford and Eileen Poiani have been wonderful team partners and have contributed

---

[2] www.maa.org/saum/cases/StPeters_B.html

tremendously to everything discussed here. The entire Saint Peter's College Department of Mathematics faculty has been supportive and engaged; they are truly dedicated to the students. The leadership and colleagues of SAUM, especially the "first class" led by Bill Marion, Bernie Madison, Bill Haver, and Bonnie Gold, have provided invaluable encouragement, commiseration, and ideas.

References

1. Poiani, Eileen. "Does Developmental Mathematics Work?" in *Assessment Practices in Undergraduate Mathematics*, edited by Bonnie Gold, Sandra Z. Keith, and William A. Marion. Washington, DC: Mathematical Association of America, 1999.
2. *Undergraduate Programs and Courses in the Mathematical Sciences: CUPM Curriculum Guide 2004*, Committee on the Undergraduate Program in Mathematics. Washington, DC: Mathematical Association of America, 2004.

Appendix A. Assessment Objectives, Courses, and Methods

Objectives and Course Table. Nine of the fifteen student learning objectives listed in Figure 2 are addressed in every major course (objectives 1, 2, 3, 6, 7, 9, 10, 12, and 15). The following table illustrates which of the remaining six objectives are addressed in particular classes. For example, Ma377, differential equations, addresses objectives 1, 2, 3, 4, 6, 7, 8, 9, 10, 12, 14, and 15.

Objective:	4	5	8	11	13	14
Course:	techno.	proof	models	research	expos.	comm.
143–144 Calculus	X		X			
246 Discrete Math		X	X	X	X	X
247 Linear Algebra	X	X	X	X	X	X
273-4 Multivariable Calculus		X	X			X
335 Probability	X	X	X			
336 Statistics			X			
375 Advanced Calculus		X				X
377 Differential Equations	X		X			X
382 Modeling	X		X	X	X	X
385 Applied Math	X		X	X	X	X
441 Modern Algebra		X			X	X

Objectives and Methods Table. The following table illustrates which of the fifteen objectives in Figure 2 are assessed by the eleven methods listed in Figure 3 (recognizing that attention to some objectives varies with instructors). For example, the column headed "k fac. talk" indicates that the assessment tool of discussions with faculty colleagues addresses objectives "9 apply", "10 transfer", "12 read", and "14 comm." (More fully, application of mathematical content to other disciplines, transference of mathematical thinking beyond cognate fields, reading and assimilating technical material, and communicating technical material at an appropriate level). Similarly, the "15 career" row shows that the objective of vocational success after completion of the major is assessed through "j survey," the surveys of graduates.

method	a	b	c	d	e	f	g	h	I	j	k
objective	class test	home test	home work	group project	comp. projecf	paper	fac. pres.	class pres.	posters	survey	fac. talk
1 think	X	X	X	X	X	X	X	X	X		
2 mature	X	X	X	X	X	X	X	X	X		
3 solve	X	X	X	X	X	X	X	X	X		
4 techno.		X	X	X	X	X	X	X	X		
5 proofs	X	X	X	X		X	X	X	X		
6 history		X	X			X	X	X	X		
7 connect.	X	X	X	X	X	X	X	X	X	X	
8 models	X	X	X	X	X	X	X	X	X	X	
9 apply				X	X	X	X	X	X	X	X
10 transfer										X	X
11 research		X	X	X	X	X	X	X	X		
12 read	X	X	X	X	X	X	X	X	X	X	X
13 expos.				X		X	X	X	X		
14 comm.	X	X	X	X	X	X	X	X	X	X	X
15 career										X	

Appendix B. Results of Transcript Analysis

	1980–1985	1986–1991	1992–1997	1998–2003	1980–2003
Majors	33	16	18	23	90
F / M*	21-Dec	10-Jun	11-Jul	11-Dec	37 / 53
API / H / W*	1 / 5 / 27	4 / 1 / 11	0 / 2 / 16	3 / 1 / 19	8 / 9 / 73
Math GPA	3.22	3.59	3.34	3.32	3.34

* Demographic data includes female/male and Asian – Pacific Island/Hispanic/White (non-Hispanic). Hispanics and blacks are significantly underrepresented in comparison to the statistics for the overall student body.

Also, 27 of these 90 graduates have gone on to complete advanced degrees (30%), including four Ph.D.s (all women).

(All subsequent tables give percentages, not actual numbers of students.)

First math course	1980–1985	1986–1991	1992–1997	1998–2003	GPA
Below calculus	3	6	0	13	3.34
Elem. calculus	21	6	28	26	3.14
Major calculus	67	69	61	39	3.36
Above calculus	9	19	11	22	3.54

Min. & 2nd Maj.	1980–1985	1986–1991	1992–1997	1998–2003
C.S.	27	63	33	34
Ed.	0	0	11	34

Prerequisites	followed / didn't	followed GPA / didn't GPA
Overall	74 / 26	3.33 / 3.34
1980–1985	70 / 30	3.24 / 3.19
1986–1991	75 / 25	3.56 / 3.67
1992–1997	94 / 6	3.37 / 2.81
1998–2003	65 / 35	3.26 / 3.44

Appendix C. Results from Alumni Survey

Course	% "very valuable"	Course	% "not valuable"
Math Modeling	85	Numerical Analysis	60
Probability	81	Modern Algebra	59
Differential Calculus	75	Modern Geometry	50
Mathematical Statistics	69	Discrete Math	46
Linear Algebra	69	Elementary Calculus	43
Elementary Statistics	67	Advanced Calculus	42
Intermediate Calculus	63	Complex Variables	40
Elementary Calculus	57	Differential Equations	38
Multivariable Calculus	56	Intermediate Calculus	25
Discrete Math	54	Multivariable Calculus	25
Pascal/C++/Other Programming	50	Pascal/C++/Other Programming	25
Differential Equations	46	Differential Calculus	17
Topics in Applied Math	43	Mathematical Statistics	15
Complex Variables	40	Topics in Applied Math	14
Advanced Calculus	33	Linear Algebra	13
Modern Geometry	25	Math Modeling	8
Modern Algebra	24	Probability	6
Numerical Analysis	20	Elementary Statistics	0

Courses Most Helpful in Career	Teachers	Non-teachers	Total
Math Modeling	2	3	5
Mathematical Statistics	2	3	5
Differential Calculus	3	1	4
Linear Algebra	3	0	3
Probability	1	2	3

Course Least Helpful in Career	Teachers	Non-teachers	Total
Modern Algebra	5	6	11
Advanced Calculus	4	3	7
Pascal/C++/Other Programming	2	2	4
Intermediate Calculus	2	1	3
Discrete Math	1	2	3

Assessing the Mathematics Major Through a Senior Seminar

Donna Flint and Daniel Kemp
Department of Mathematics and Statistics
South Dakota State University
Brookings, SD
donna_flint@sdstate.edu,
daniel_kemp@sdstate.edu

Abstract. At South Dakota State University (SDSU), assessment of the Mathematics Program is achieved during the Senior Seminar (capstone) course. In this course, students participate in activities to develop their communication skills (both oral and written), prepare a portfolio, write a Major Paper, and give a Major Presentation. We describe the Senior Seminar course and the assessment tools used, and give data based on three semesters of assessment in the course. We also discuss pitfalls of the current assessment system and changes which have occurred because of information obtained through the assessment process.

South Dakota State University (SDSU) is the largest university in a regional system of six universities in South Dakota. Enrollment over the last five years has ranged from 8,000 to 10,000 students. The Department of Mathematics and Statistics is in the College of Engineering. The number of mathematics majors has varied over the last few years from 60–75 students, with approximately 15 majors graduating each year. SDSU also has a Master's program which has ranged from 10–16 students enrolled per year. The department currently has 17 full time faculty.

Background

Beginning in the 1980s, in addition to regular coursework, students were required to read 15 articles from mathematics journals and write reaction papers for each article. In the early 1990s, a departmental competency test, with no grade attached, was added to requirements for graduation. In 1996, in response to the university assessment requirement, a senior seminar (capstone) course replaced both the reading and competency test requirement. The main activity of this course was to have students write a major paper and give a presentation based on that paper. In 2002, in order to better align with departmental and university mission statements as well as the department learning objectives, and to more accurately collect assessment information, Senior Seminar was revamped. In addition to a major paper and presentation, students engage in activities that give them experience researching, writing, and presenting. At this time, the Senior Seminar instructors along with the Department Head formed the Assessment Committee which reports regularly to the University Assessment Committee. The course is still in constant revision, and in fact beginning in the Fall of 2004, the course will be expanded to be a two semester, two credit course.

Departmental Goals

The following are student attainment goals taken from the Department Standards Documents. These goals are assessed throughout the student's academic career (demonstration of competence is documented when a student earns at least a "C" in mathematics courses) as well as in the requirements for Senior Seminar.

1. Demonstrate competence in all core areas of undergraduate mathematics.
2. Use contemporary mathematical and presentation software and technology.
3. Apply research methods to mathematical problems.
4. Communicate clearly and succinctly in writing in the discipline.

5. Articulate complex ideas to an audience.

6. Reflect on learning experiences over an extended period of time period to identify areas for further learning.

Senior Seminar as an assessment tool

Senior Seminar is required of all graduating mathematics majors. Majors take it in their final semester on campus (education students must take it the semester before they student teach). An example of a sample schedule for this one-semester course is found in Appendix A. Starting in the Fall of 2004, majors will take it in the last two semesters they plan to be on campus. Since the degree earned by our students is a Bachelor of Science in Mathematics (even if they are earning their teaching certification), the emphasis in senior seminar is in pure and applied mathematics. Education issues are addressed in the math education courses. The major focus of Senior Seminar is to help the students develop the skills needed to write a solid mathematics expository paper and make a presentation based on this paper. During the first semester, students will participate in several activities (we will discuss these later), choose a topic for their major paper and start research for this topic. During the second semester, students will continue to work with their paper advisors to research, write, revise and present their papers. At the conclusion of the course, students will also submit a portfolio which documents their academic career at SDSU. Contents of this portfolio will be discussed later. Prior to Fall of 2004, all of this was completed in one semester.

Each student paper, presentation and portfolio is assessed by faculty members. The data obtained from these assessments are the groundwork for assessing the mathematics major at SDSU. Though transcript data is necessarily a part of assessment, it is not included as part of our assessment of the major. The understanding is that students will not be considered for graduation unless they have satisfied all course requirements.

First semester Senior Seminar activities

The first semester of Senior Seminar is a time for students to develop some of the skills which will enable them to research and prepare a major paper and presentation. It is also used to assess some of the learning objectives for the major.

To develop research skills:

- History quiz: Students are given a list of important historic mathematical events and they are to determine the dates, people involved and significance of the event.

- Journal article: Students are assigned a journal article to read. They are given the source and can find the article either in the University library or by using an on-line source. They are to read the article and write a paper discussing their thoughts on the topic involved. Sometimes this assignment will also include some mathematics problems to be solved.

- Web research project: Students choose one of two possible projects. If they plan to go to graduate school, they research a math field, describe the field, how it started, people who originated the field, and who is working in that field now. Other students research a math career and write a short paper outlining what is involved in the career, how to prepare, prospects for the career, and what kind of growth they might expect in that career.

- Web Research project: Students are to find a proof of some non-trivial mathematical statement on the web. They are to read and understand the proof, verify that it is valid, then rewrite the proof in their own words.

To develop communication (oral and written) skills:

- History quiz: Students present findings from the history quiz to the class and also write a short paragraph for each event assigned.

- Calculus problem: Students are given a multi-faceted problem which can be solved using calculus. Students solve the problem, write their solution (including proper notation and appropriate diagrams) using mathematical software, then present the solution of their problem to the class.

- Faculty Talks: Faculty members from SDSU and invited speakers make presentations to the senior seminar class. Presentations are generally on topics the students have not yet seen. Students learn about varied fields of mathematics, as well as observe some good and bad presentation techniques. Students are to write reaction papers discussing their thoughts on the topic and the presentation techniques.

Assigning Grades:

- The class assignments count as 25% of the student's final grade. Each is weighted depending on the amount of work required to complete the assignment. Grades are very subjective, and reflect the mathematical accuracy of the work, where applicable, but more importantly reflect the student's ability to communicate effectively.

The Major Paper

The major paper is an expository paper in which the student explains a mathematical topic. This paper counts as 35% of

Categories for Assessing Mathematics Majors' Papers	Assessment Criteria
Presentation: 1. How well written is the report (e.g., correct grammar, spelling, etc.)? [4] 2. How effectively does the student use graphics/figures (e.g. placement, neatness, etc.)? [2,4] 3. How effectively does the student use examples to clarify points made in the paper? [4] **Technical Report:** 4. How well does the student explain the purpose of the project? [4] 5. How well does the student explain the history of the topic? [3,4] 6. How well does the student explain the background mathematics needed to understand the problem? [3,4] 7. How well does the student use and cite both print and electronic sources? [3] **Mathematics/Statistics** 8. How in depth does the student explore the topic? [1,3] 9. How accurate are the mathematical/statistical statements? [1,3] 10. How well does the student justify the mathematical/statistical statements? [1,3,4] 11. How consistent and effective is the student's use of notation? [1,3,4] **Overall** 12. How well does the student demonstrate that he/she understands the mathematics/statistics? *(Learning objectives are listed in brackets.)*	7 – Excellent—has no mathematical errors may need some minor re-wording 6 – Very Good—needs minor revisions in a few places 5 – Good—needs minor revisions sporadically throughout the paper 4 – Satisfactory—minor revisions needed throughout the paper 3 – Poor—needs major revisions in a few places 2 – Very Poor—needs major revisions sporadically throughout the paper 1 – Unsatisfactory—paper has little or no merit

Figure 1. Criteria for assessing senior mathematics majors' research papers.

the student's course grade. The paper contains "significant mathematics" and at least one major proof. It is word processed, with appropriate mathematical symbols and diagrams. The body of the paper is eight to twelve pages. Students are expected to follow a prescribed format which includes title page, abstract, bibliography, and biographical statement. The paper is based on a topic chosen by the student in consultation with a mathematics faculty member who supervises the paper. Students are required to meet once a week with their advisor in order to discuss and ensure progress on the paper throughout the semester.

The topics of the papers vary greatly. Some students integrate their paper with some outside interest (for example, math and music, mathematics of castle defense, statistical model of a baseball park) while others choose a purely mathematical topic (for example, circle inversion and the shoemaker's knife, cubic equations, linear programming, differential equations involving repeated eigenvalues). The most important aspect of the paper is that the topic is beyond the student's coursework. The student gathers sources, learns about the topic, and writes a summary of information gathered and learned. This work is done independently. The faculty member's role is that of guidance—the student does not learn the material primarily from his or her advisor. Emphasis is both on learning new material as well as proper written presentation of a paper- including proper formatting and notation, citation of sources, and good mathematical exposition.

The major paper is assessed by at least two faculty members, and the papers are judged on physical presentation, technical preciseness and the student's demonstration of their understanding of the mathematics involved. (See the Insights section below for a discussion of evaluator consistency.) Assessment categories and criteria are shown in Figure 1; a copy of the actual assessment form can be found in Appendix B of our expanded case study on the SAUM website.[1]

The major presentation

The major presentation is based on the major paper described above. The presentation counts as 25% of the student's course grade. Each student prepares a Microsoft PowerPoint presentation based on a portion of her/his major paper. This is delivered to their peers in Senior Seminar as well as to faculty who assess the presentation. The presentation is about fifteen minutes long with an additional ten minutes allowed for questions. Due to time limitations, students are advised to select only a portion of their paper for the presentation. Many students include other technology tools in their presentation such as Geometer's Sketchpad, Maple, and applets downloaded from the Web. Students are evaluated on both their ability to engage the audience with an interesting presentation and their demonstration of understanding of their chosen topic.

---

[1] www.maa.org/saum/cases/SDakota_A.html

Categories for Assessing Major Presentations	Assessment Criteria
Delivery 1. How well did the student explain the purpose of the project? [5] 2. How well did the student organize the material? [5] 3. How well did the student's presentation make use of overheads, PowerPoint, and/or other technology? [2] 4. How would you judge the student's presence? (voice, delivery, etc.)? [5] 5. How well did the student answer/respond to questions and comments? [5] **Content** 6. How well did the student communicate his/her understanding of the mathematics? [1,3,5] 7. How well was the mathematics developed through logical presentation, justification of assumptions and examples? [3,5] **Overall** 8. How well did the student demonstrate that he/she understands the mathematics/statistics?	7 – Excellent—ready for presentation at a conference 6 – Very Good—needs refinement before presenting at a conference 5 – Good—needs 1–2 minor changes before presenting at a conference 4 – Satisfactory—needs 3–5 minor changes before presenting at a conference 3 – Poor—needs 1–2 major changes before presenting at a conference 2 – Very Poor— needs 3–5 major changes before presenting at a conference 1 – Unsatisfactory—presentation has little or no merit

Figure 2. Criteria for assessing major presentations.

The major presentation is assessed by at least three faculty members, but can be assessed by any faculty or graduate teaching assistant who chooses to attend the presentation. The presentations are judged on delivery, content and overall presentation. Assessment categories and criteria are shown in Figure 2; a copy of the actual assessment form can be found in Appendix C of our expanded case study on the SAUM website.[2]

The portfolio

The Portfolio is turned in at the end of the student's final semester at SDSU and counts as 15% of the student's course grade. The students are given a copy of portfolio requirements in the Logic and Set Theory course; this is the first course required of all Mathematics Majors. This would normally happen during the Sophomore year, but there are always exceptions. The portfolio is used as an assessment tool by the Senior Seminar Instructors, used by the Department Chairman to guide the Exit Interview, and used by the Students when they begin searching for a job. In the portfolio, students include documents which demonstrate competency in five mathematics courses. Included with each of these submissions is a paragraph explaining how each item included demonstrates competence. In addition, students include their major paper, a resume, and several essays reflecting on their mathematics career so far and the mathematics program at SDSU. Specific requirements and the assessment criteria can be found in Appendix B at the conclusion of this paper.

Insights: Thoughts on the data and the cycle of assessment

A summary of the data obtained can be found in Appendix C. The data has been drawn from three semesters of data collection. Though Senior Seminar has been in place for several years, we only began collecting numerical data in the Fall of 2002.

First, we feel the data we have obtained is not necessarily consistent. Suggestions for normalizing these evaluations are to make the scale smaller, train evaluators and to give examples of what should be graded as 1, 2, etc. We are currently rewriting the evaluation criteria and hope that this will enable the faculty evaluators to be more consistent in grading. An example of the new format can be found in Appendix D. Because of this inconsistency, it is hard to justify changes based only on the data obtained. Therefore, many changes we have made are based on observations made by faculty members and students.

One major change brought about by the assessment cycle is the addition of a second semester of Senior Seminar. This change came about based on information in the Student Portfolios, as well as evaluation of papers and presentations. A second semester gives the student more time to practice writing in the first semester and more time to actually do their writing in the second semester. This also gives the student more opportunity to evaluate oral presentations in the first semester and more time to prepare their own presentation in the second semester. The hope is that this will result in better quality papers and presentations.

Other changes that included a revamping of the Mathematical Applications in Computers course which

---

[2] www.maa.org/saum/cases/SDakota_B.html

came about because of feedback found in the Student Portfolios. Students commented in their department evaluation that they did not feel they worked with enough software in the course, that the emphasis was on linear algebra, not computers, and that the 7:00 AM meeting time was too early. Therefore, the revamping included more computer use, more variety of topics, updated software, and a time change. Based on assessment of the Major Paper and Presentation, more specific guidelines in the topic choice and interaction requirements with the paper advisor as well as more specific guidelines for writing the Major Paper and giving the Presentation were outlined. In addition, based on feedback from all three sources, more frequent deadlines were arranged to better monitor progress on the Major Paper. We found that involvement of more faculty in the assessment process made more faculty aware of the need to improve student Papers and Presentations.

Acknowledgements. We would like to thank the MAA, NSF, and organizers of the SAUM Workshops for the opportunity explore assessment with guidance and with peers. We also appreciate new ideas suggested by participants in the SAUM workshop. We also thank members of the SDSU Department of Mathematics and Statistics for their encouragement and participation in our assessment activities.

Appendix A. Senior Seminar Class Schedule (Fall 2003)

(Each entry below represents one class period; the class meets once a week for 50 minutes.)

Plan: Math History Quiz

Assignment: Journal article: read, write reaction paper, be prepared for discussion "A Discrete Look at $1+2+\cdots+n$" by Loren C. Larson

Due Today: Paper Topic and Advisor and Math History Quiz

Plan: Discuss History Quiz and Calculus problem

Assignment: Write-up of the Calculus problem (hand-written)

Due Today: Hand-written write-up of Calculus problem

Plan: Computer lab: Learn Scientific Notebook , MS Word, WinPlot to write-up Calculus problem

Assignment: Complete word processed write-up of Calculus problem including the mathematics

Due Today: Reaction paper for Journal Article: Three sources for major paper, one from the web

Plan: Student led discussion based upon Journal Article

Due Today: Word Processed outline of major paper/ Word Processed write-up of Calculus problem

Plan: Student presentations of outlines of papers- peer comment

Due Today: Word processed bibliography for major paper

Plan: Complete student presentations.

Assignment: Write a resumé (or update your current resumé)

Due Today: Word Processed "very rough draft" of major paper

Plan: Resumé discussion and Computer lab web research

Assignment: Write a 1–2 page paper describing career or field of mathematics

Due Today: Assigned paper

Plan: Faculty talk: Assignment: Reaction paper/ assignment

Due Today: Math career or math field paper

Plan: Faculty talk: Assignment: Reaction paper/ assignment

Due Today: Draft of major paper submitted to your advisor and Reaction paper/assignment

Plan: Faculty talk: Assignment: Reaction paper/ assignment

Advisor comments due to student and senior seminar advisors

Due Today: MAJOR PAPER and Reaction paper/assignment

Plan: Faculty talk: Assignment: Reaction paper/ assignment

Due Today: Reaction paper/assignment

Plan: Faculty talk: Assignment: Reaction paper/ assignment

Plan: MAJOR PRESENTATIONS (Evening)

Due today: Reaction paper/assignment

Plan: Program Evaluation

Final Exam Day — Portfolio is due by 5:00

Appendix B. Portfolio Requirements

Student: _____   Faculty Reviewer: _____

This portfolio will be assessed on selection of material (how well it satisfies criteria) and your reflection upon that material. Each item will be given a rating on a scale. The numbers in brackets represent the departmental goal measured by this criteria.

> 4 – excellent (good choice/thoughts insightful),
>
> 3 – good (good choice/needs more reflection)
>
> 2 – poor (poor choice/very little or no reflection)
>
> 1 – not included

A. Materials from the Mathematics courses you have taken that illustrate the following, with paragraphs for each explaining how what you have chosen demonstrates your competency:

    i.   Competence in Calculus   _____   [1]

        Competence in Linear Algebra   _____ [1]

        Competence in upper level course   _____ [1]

        Competence in upper level course   _____ [1]

        Competence in upper level course   _____ [1]

    ii.   Ability to write a clear and correct proof   _____ [1,4]

    iii.   Use of mathematical software   _____ [2]

    iv.   Example of a test on which you did well; discuss why you think you did well   _____ [6]

    v.   Example of a test on which you did poorly; discuss what you think you did wrong in preparing for this test   _____ [6]

B. Your revised Senior Seminar Research Paper   _____ [1,3,4]

C. A current Resume   _____ [6]

D. A letter of application for a job   _____ [6]

E. A written summary of general University extra curricular activities   _____ [6]

F. A written evaluation of the mathematics program at SDSU   _____ [6]

G. An essay reflecting upon your career as a student in Mathematics   _____ [6]

H. An essay written in your sophomore year discussing your plans for your academic career.   _____ [6]

I. Other materials:

    Mathematics Majors with teaching certification:

        An essay on your student teaching experience   _____ [6]

        A letter from your cooperating teacher (not graded)

        Report of your grade on the PRAXIS exam   _____ [1]

    Mathematics Majors without teaching certification

        An essay on how your mathematics training will be used after graduation   _____ [6]

                                      Total: _____ / 68 or 72 (ed)

J. Additional materials: maximum of 5

                                      Total: _____ / _____

Appendix C. Three Semesters' Data

The data below has been compiled from Fall 2002 (4 students), Spring 2003 (2 students), and Fall 2003 (10 students).

Major Paper (Scale is 1-7):

	Average Score Based on Faculty Responses		
Learning Objective:	Fall 2002	Spring 2003	Fall 2003
1 (competence)	5.47	4.97	5.59
2 (technology)	5.64	6.00	5.52
3 (apply research methods)	5.50	5.18	5.67
4 (communicate in writing)	5.54	5.39	5.67
5 (communicate verbally)	N/A	N/A	N/A
6 (reflect on learning experiences)	N/A	N/A	N/A

Major Presentation (Scale is 1-7):

	Average Score Based on Faculty Responses		
Learning Objective:	Fall 2002	Spring 2003	Fall 2003
1 (competence)	5.81	4.18	5.1
2 (technology)	6.10	6.64	5.65
3 (apply research methods)	5.71	4.32	5.16
4 (communicate in writing)	N/A	N/A	N/A
5 (communicate verbally)	5.92	5.32	5.55
6 (reflect on learning experiences)	N/A	N/A	N/A

Portfolio (Scale is 1-4):

	Average Score Based on Faculty Responses		
Learning Objective:	Fall 2002	Spring 2003	Fall 2003
1 (competence)	3.69	3.958	3.917
2 (technology)	4	4	4
3 (apply research methods)	4	3.5	3.5
4 (communicate in writing)	3.76	3.833	3.813
5 (communicate verbally)	N/A	N/A	N/A
6 (reflect on learning experiences)	3.78	3.958	3.917

Appendix D. Revamped Assessment Tools (Sample)

Student: _____   Faculty Reviewer: _____

Overall Paper grade: _____

You are assessing a senior Mathematics Major student's research paper. This paper was written with the supervision of a Mathematics Faculty member. You should assess the paper based on the following criteria. Please circle the number which best describes the paper. Please note the even numbers can be used to show that the paper shows qualities of both neighboring categories (feel free to circle relevant characteristics). Learning objectives are listed in brackets. Please add additional comments for the student on a separate sheet of paper. Finally, assign an overall grade (A, B, C, D, F) to the paper. Students will only see averages of all readers and comments, but will not know the names of readers.

1. How well written is the report (e.g., correct grammar, spelling, etc.)?   [4]

1	2	3	4	5	6	7
Many grammatical or spelling errors Frequent misuse of mathematical language Use of Slang		Periodic grammatical or spelling errors Periodic misuse of mathematical language Periodic use of slang		Few or no grammatical or spelling errors Some misuse of mathematical language Slang mostly avoided		No grammatical or spelling errors Precise mathematical language used Slang avoided

2. How effectively does the student use graphics/figures (e.g. placement, neatness, etc.)?   [2,4]

1	2	3	4	5	6	7
Many missed opportunities for necessary graphics Graphics included are sloppy, unnecessary, and/or unjustified		Occasional missed opportunity for appropriate graphics Not labeled correctly Inclusion of the graphics not justified in the text		Graphics used where appropriate Labeled correctly Inclusion of some graphics not well justified in the text		Graphics used where appropriate Labeled correctly Easy to read Inclusion of graphics justified in the text.

3. How effectively does the student use examples to clarify points made in the paper?   [4]

1	2	3	4	5	6	7
Many missed opportunities for necessary examples Trivial or unimportant examples included Examples not explained well		Occasional missed opportunity for appropriate examples Unnecessary examples included Examples not thoroughly explained nor accurate		Examples used where appropriate Examples explained clearly Some examples not necessarily relevant Some examples not necessarily accurate		Examples used where appropriate Examples explained clearly and accurately

4. How well does the student explain the purpose of the project?   [4]

1	2	3	4	5	6	7
Purpose not explained Purpose not reasonable Not explained in a reasonable part of the paper		Purpose explained in a cursory manner Purpose is contrived Not explained in a reasonable part of the paper		Purpose explained reasonably well Purpose is reasonable Purpose explained in appropriate part of the paper		Purpose explained thoroughly Purpose is reasonable Purpose explained in appropriate part of the paper

Assessment of Bachelors' Degree Programs

James R. Fulmer and Thomas C. McMillan
Department of Mathematics & Statistics
University of Arkansas
Little Rock, AK
jrfulmer@ualr.edu, tcmcmillan@ualr.edu

Abstract. In this case study, we will examine the process that we currently use for assessing the baccalaureate degree programs in the Department of Mathematics and Statistics at the University of Arkansas at Little Rock (UALR). Over the last year, the authors of this study have participated in the MAA workshop Supporting Assessment of Undergraduate Mathematics (SAUM). We will include details of how insights gained at this workshop have been incorporated into our assessment of bachelor's degree programs.

Assessment of Mathematics at UALR

The assessment process at UALR contains several components. Each degree program has an assessment plan, which describes how that program is assessed each year. The assessment cycle covers the calendar year, January 1 through December 31. During this time, various assessment activities are conducted to collect the data prescribed by the assessment plan. In January, each program prepares an Assessment Progress Report, covering the previous year. The report should focus on 1) the use of assessment for program building and improvement, 2) the faculty and stakeholder involvement, and 3) the methods defined by the assessment plan. These reports are evaluated by the College Assessment Committee, using a rating scale of 0 through 4, on the basis of the three items previously listed. The College Assessment Committee compiles a College Summary Report and submits it to the Dean of the college in March. All assessment reports are due in the Provost's Office by April 1. The chairs of the college assessment committees form the Provost's Advisory Assessment Group. This committee meets monthly and establishes overall policies and guidance for program assessment on campus.

The Department of Mathematics & Statistics at the University of Arkansas at Little Rock has three ongoing assessment programs: core assessment, undergraduate degree assessment, and graduate assessment. This study deals only with the undergraduate assessment program. At the time we entered the SAUM workshop, our department had already designed and implemented an assessment process. Our experience with the process identified its shortcomings and our participation in SAUM gave us insights into how the process could be improved. What has resulted is not so much a new assessment process, but a logical restructuring of the existing process so that more meaningful data can be collected and that data can be interpreted more easily. Since most of the instruments that we use to collect data were in use before we implemented the changes, in this paper we will concentrate on the new logical structure of our assessment process and give only a brief description of the problems and shortcomings that we identified in the earlier assessment program.

The main problem with the process used to assess undergraduate degree programs was that the data being collected were only loosely related to departmental goals and student learning objectives. Our department has established a mission statement, goals and student learning objectives. However, the data collected from student portfolios, student presentations, alumni and employer surveys, and the exit examination were not easily interpreted in a way that measured our relative success in achieving these goals and objectives. Another problem we encountered is the low return rate for alumni and employer surveys. Finally, we found that, although we seemed to have a massive amount of assessment data, there would be so few data points relating to a particular student learning objective as to be statistically insignificant. The result of the assessment process was an annual report that beautifully summarized the data we collected, but did not clearly suggest trends. The difficulty in interpreting the data presented an impediment to the successful completion of the most important part of the assessment cycle: using the result of assessment to improve the degree programs.

New Directions in Assessment at UALR

Assessment in the Department of Mathematics and Statistics continues to be driven by the goal statement that is published in the university catalog:

"The objectives of the department are to prepare students *to enter graduate scho*ol, to teach at the elementary and *secondary levels, to understand and use mathematics in other fields of knowledge* with basic mathematical skills for everyday living, and *to be employed and to act in a consulting capacity on matters concerning mathematics*." (Emphasis added to identify items in the department's mission statement that are relevant to baccalaureate degree assessment.)

Using insights we gained in SAUM, we have given our assessment process a logical structure that should make interpretation of the data more natural. We have redesigned the logical structure of assessment using a "top-down" approach. The department has identified several student learning objectives that are solid evidence of our students' meeting the department's established goals. For each of these student learning objectives, we established "assessment criteria", which, if satisfied by the students, are strong evidence that the objective has been attained. Finally, for each assessment criterion, we established one or more "assessment methods" for gathering evidence that the students have satisfied the criterion. The top-down approach to assessment that we developed over the year of our participation in SAUM is summarized in Figure 1.

This top-down approach has two significant advantages. First, since each assessment method is explicitly related to a set of assessment criteria, assessment instruments can be designed to collect the best possible data for measuring student achievement on that criterion. Here is an example. One of our assessment criteria is that students should be able to

demonstrate at least one relationship between two different branches of mathematics. We have looked for evidence for this criterion in student portfolios, where it may or may not have been found. Under our new scheme, since we anticipate that student portfolios will be used to evaluate this criterion, the process for completing portfolios has been redesigned to guarantee that portfolios contain assignments in which students attempt to demonstrate a relationship between two different branches of mathematics. Students in the differential equations course, for example, can be given a portfolio project that draws on their knowledge of linear algebra. Students in advanced calculus may be asked to draw on their knowledge of geometry or topology.

The second advantage of this top-down approach is that sufficient data will be collected relative to each assessment criterion. The assessment process involves the independent evaluation of student work (portfolios, written and oral presentations) by members of the department's assessment committee. Each committee member is guided in his or her evaluation by a rubric in which each question has been specifically designed to collect data relating to an assessment criterion. The design of all assessment instruments (including surveys, rubrics and interviews) is guided by the assessment criterion they will measure. The explicit connection between assessment method and assessment criterion will facilitate the interpretation of the data. Although it may not be clear in the first few assessment cycles whether the data suggest a modification of the assessment method or an improvement in the degree program, it is evident that convergence to a meaningful assessment program, which provides useful feedback, will not occur if this explicit connection between assessment criterion and assessment method is not made.

Mathematics and Statistics faculty are responsible for collecting and interpreting assessment data. The department coordinates its assessment activities with the college and university. The next to the last step in the assessment process at UALR is the preparation of an assessment progress report that is evaluated by our colleagues in the College of Science and Mathematics. The assessment progress report is made available to all interested faculty at the annual College Assessment Poster Session. Every assessed program is represented at this spring event with a poster that summarizes the results included in the report. The critical final step of our new process will be a departmental assessment event at which faculty members give careful consideration to the report prepared by the Departmental Assessment Committee and the evaluation from the College Assessment Committee. This most important step is the "closing of the feedback loop." All mathematics faculty will examine the assessment

Learning Objective	Assessment Criterion	Assessment Method
Mathematics majors develop an appreciation of the variety of mathematical areas and their interrelations.	Students should be able to name several different fields of mathematics they have studied.	Senior seminar exit interview
	Students should demonstrate at least one relationship between different mathematical fields.	Portfolio review Senior seminar exit interview
Mathematics majors acquire the mathematical knowledge and skills necessary for success in their program or career.	Students should achieve an acceptable score on a nationally recognized test with comparisons to national percentiles	ETS Major Field Test
	Students should be confident that they have acquired sufficient knowledge and skills for their chosen careers in mathematics.	Alumni/student survey
Mathematics majors develop the ability to read, discuss, write, and speak about mathematics.	Students should make a presentation to their peers, including department faculty	Senior seminar final project
Mathematics majors develop the ability to work both independently and collaboratively on mathematical problems	Students should, working on their own, demonstrate the ability to solve a variety of mathematics problems.	Portfolio review Employer survey
	Students should, working collaboratively in a team setting, demonstrate the ability to solve a variety of mathematical problems.	Senior seminar Employer survey
Mathematics majors develop an appreciation for the roles of intuition, formalization, and proof in mathematics.	Students show that they can reason both intuitively and rigorously.	Portfolio review Senior seminar
	Students will show that they can reason both inductively and deductively.	Portfolio review
Mathematics majors develop problem solving skills.	Students will show they have problem solving skills.	Portfolio review ETS Major Field Test Employer survey Alumni/student survey

Figure 1. A "top-down" approach to assessment.

data for evidence that suggests appropriate changes to the degree program.

Schedule of assessment activities

Collection of data for assessment at UALR covers the calendar year, January through December. Assessment activities cover a four semester cycle: spring, summer, fall, and a follow-up spring semester. The following schedule describes these assessment activities.

Early in the spring semester, the department assessment committee identifies about five or six courses as "portfolio courses" for assessment purposes during the calendar year. The instructor of a "portfolio course" is responsible for making assignments for students that will gather information pertaining to the student learning objectives in our assessment plan. The instructor collects these "portfolio assignments" at the end of the semester and places them in the students' portfolios. Here are some examples of portfolio assignments:

- "Everywhere continuous and nowhere differentiable functions" (Advanced Calculus). Students survey mathematics literature for examples of functions continuous at every point and differentiable at no point.
- "Measure theory" (Advanced Calculus). Students explore the concept of measure theory, including Lebesgue measure, and the connections with integration theory.
- "Mixing of solutions by flow through interconnected tanks" (Differential Equations). Students explore, using a system of differential equations, the asymptotic mixing behavior of a series of interconnected tanks with inputs from a variety of sources and output to a variety of destinations.

A second assessment activity is Mathematics Senior Seminar/Capstone course in which students enroll during the spring of their senior year. One of the requirements of the course is the ETS Major Field Test, which is required of all majors in the baccalaureate degree program. We also strongly urge students in the baccalaureate mathematics degree programs to take the ETS Major Field test during their junior year. Thus, we can accumulate data on how students improve between their junior and senior year with regard to scores on the ETS MFT mathematics test. On the advice of ETS, we have not established a cut-off or passing score that mathematics majors must make in order to graduate or pass the senior seminar course. We, of course, want our students to give their best efforts on the ETS MFT. One incentive is a departmental award for the student(s) who score highest on the examination. We also appeal to students sense of citizenship in the department ("Your best effort will help us improve the program and will benefit students who follow you.") Finally, students are aware that their scores on the MFT are a part of their record within the department and will be one factor in how professors remember them.

A third assessment activity is an oral presentation made by each student to peers and mathematics faculty during the Senior Seminar/Capstone course. This presentation is based on a project that the student has developed during the senior seminar course. The oral presentation is to be supported by a written handout report describing its details. The oral presentation and written reports are evaluated by faculty using rubrics that have been designed to collect data for measuring the assessment criteria. A fourth assessment activity during the senior seminar/capstone course for each major is an exit survey, administered near the end of the course. The survey includes both subjective and objective response questions.

During the summer semester, the department assessment committee evaluates the portfolios, which now contain the spring portfolio assignments of each mathematics major, using a portfolio rubric that was developed by the department faculty. Instructors of "portfolio courses" that had been designated early in the spring semester, continue to make and collect certain "portfolio assignments" that provide data for measuring the student learning objectives.

During the fall semester, instructors of portfolio courses continue making and collecting certain portfolio assignments. A second activity is administering the alumni and employer surveys. Both surveys are sent by mail to each alumnus with the instruction that the alumnus is to pass along the employer survey to his or her employer. Self-addressed, postage-paid envelopes are enclosed in order to facilitate and encourage a response from each alumni and employer. The assessment activities of the fall semester complete the calendar year of collecting data for assessment purposes.

During the follow-up spring semester, the department assessment committee begins the process of evaluating assessment data collected during the previous calendar year. The department assessment committee meets and evaluates

the latest additions to the portfolios. The committee then writes the assessment progress report, which is due on March 1 of each year. In writing this report, the committee considers the scores on the ETS-MFT test, student portfolios, faculty evaluations of the students' oral and written reports, exit surveys for majors, alumni surveys, and employer surveys. This data is evaluated with respect to the assessment criteria with the goal of measuring how well the student learning objectives have been met. All of this goes into writing the assessment progress report. A College Assessment Poster Session, where a summary of the assessment progress report is displayed on a poster, is held during March.

The assessment progress reports are collected by the College Assessment Committee, consisting of one member from each department in the college. The College Assessment Committee is divided in teams of two each to evaluate the department assessment progress reports. Each team of two is selected so that at least one member served on the committee the previous year and is a continuing member; also, the team is selected so that no member of the team is from the department whose assessment progress report is being evaluated. The team evaluates the assessment progress report with a scoring rubric that is used campus-wide. The department assessment committee then considers the assessment evaluation report and all other assessment data collected during the calendar year and analyzes how well the student learning objectives are being met. It is at this point in the process that the department will make data-driven decisions concerning possible changes to the mathematics curriculum. This completes the most important part of the assessment cycle, "closing the loop" by using the results of assessment to improve the degree programs.

Conclusions

This case study should be considered a preliminary report. The changes to the structure of our assessment program were made during the year of our participation in SAUM. The evaluation of the newly restructured assessment cycle will not be completed until spring, 2003. A preliminary examination of our collected data has given us confidence that our assessment process has been significantly improved. For example, we have now collected faculty reviews of student portfolios. There is now an explicit link, via the inclusion of assessment criteria in the evaluation rubrics, between the data that comes from these evaluations and our learning objectives. The changes in the logical structure of our assessment process were motivated by the shortcomings that we recognized and the very good advice that we got from our colleagues and mentors in SAUM.

Acknowledgements. We are both very appreciative of the MAA support that enabled us to participate in SAUM. The great value of our interactions with colleagues at other colleges and universities who face similar problems is difficult to measure. We are especially appreciative of our mentor, Bill Marion, whose suggestions have resulted in significant improvements to our assessment process. We also thank Bernie Madison, Bill Haver, and Bonnie Gold for the many informative sessions that they had with us. Finally, we thank Peter Ewell for his very instructive consultation.

Assessment of the Undergraduate Major without Faculty Buy-in

Edward C. Keppelmann
Department of Mathematics and Statistics
University of Nevada
Reno, NV
keppelma@unr.edu

Abstract. In response to accreditation requirements, UNR administration mandated that we assess our major. No consensus on how to proceed could be reached. In order to avoid losing operating money, the chair had to assess anyway. What worked well was to look for prominent problems. The department chair devised a simple spreadsheet tracking system and formalized a pre-existing exit interview scheme. The main problem that emerged was a high failure rate in our key analysis sequence. Our investigations showed that this was the result of inconsistent instruction and some misconceptions about the course. By promoting discussion it was possible to simultaneously engage the faculty and satisfy the administration. Important lessons to help others in handling similar situations are summarized at the end.

Background

In addition to numerous temporary instructors the department has 24 regular faculty. We offer bachelor of science and master of science degrees with options in mathematics, applied mathematics and statistics. In response to an accreditation visit, administrators decided that every department on campus must assess its majors. Unlike other useful assessments that math departments could do, such as for example an analysis of their lower division core liberal arts offerings or their service courses for engineers, looking at our majors required getting all faculty on board. Since departmental governance is highly democratic, all major policy decisions require a vote of the regular faculty. For assessment of the major, none of the standard approaches such as portfolios (which were seen as too much work) or exit exams (which were seen as bad for recruitment of majors) could garner a majority of support and thus reaching a consensus on how to proceed became impossible.

Details of the assessment

With no other choice but to assess alone, the chair settled on a very simple spreadsheet-based tracking system. This showed the courses and the grades received for each student. This did reveal some useful patterns. In particular, performance in Analysis I and II were observed as key predictors of overall success. In addition, failure rates were high and variable from year to year.

> **Departmental Mission Statement**
> The undergraduate mathematics and statistics major seeks to develop students who have knowledge in a broad set of content areas. These will vary in depth and precise content with program option but will include a significant number of the following:
> - Calculus (Required)
> - Linear Algebra (Required)
> - Differential Equations (both PDE and ODE)
> - Analysis (proofs-oriented advanced calculus) (Required)
> - Probability and Statistics
> - Numerical Methods
> - Discrete Mathematics (graph theory, combinatorics, game theory)
> - Topology
> - Abstract Algebra
> - Complex Variables
> - Mathematical Modeling (although largely interdisciplinary with many of the other topics listed above, we also teach the difference in philosophy and approach required for many applied problems.)
>
> In the process of obtaining this knowledge, the student will also achieve a significant number of learning outcomes described in Appendix A.

Long before the assessment mandate, the previous chair used a program of exit interviews with graduating seniors. Although this has been entirely the chair's responsibility, and thus a significant burden, it is a very useful activity. By providing a unique global perspective of department faculty, such a scheme is important both for assessment and annual merit evaluations. Because of the mandate, we decided to formalize this process. (See Appendix B for our questions).

In exit interviews it can be hard to get honest answers to questions like "Who was your least effective instructor?" or "Do you have complaints about our program?" Despite this, some very useful ideas for improvement were obtained but it wasn't always clear how to use these insights. For example, some professors were identified as inadequate in fundamental ways. How seriously should teaching assignments be reconsidered based on this?

As with any exit interview system, accuracy in measuring learning outcomes is severely limited. For example, questions that ask about the mastery of specific academic topics cannot be fully trusted. Most students simply do not possess the maturity to properly judge their level of expertise or the precise role of various subjects within their future needs. In many cases, however, the interview can be a good learning experience for students (especially when conducted in groups) and a recruiting tool for the department's graduate programs.

Findings

In addition to the variable passing rates and success which was highly dependent on performance in our analysis sequence, grade tracking also revealed a large increase in enrollment in Analysis I. Class sizes, which just a few years earlier had been in the low teens and below, were now regularly above 30. We were puzzled that there was no corresponding rise in graduation rates.

Exit interview data partially confirmed these observations. In all, 16 of 22 interviewed students reported that Analysis I & II was a massive culture shock that many students simply could not recover from. This is the transition between computational mathematics and the deeper theoretical aspects of the subject. Many seniors reported that other students changed their major as a result of Analysis.

These findings suggested that a more detailed examination of the courses was needed. Consequently, the chair focused closely on the tracking of all students (not just majors) who had taken Analysis I in the years from 1998-2001. He chose those years since virtually all the students involved would have completed or dropped out by now.

Furthermore, the instructor was different in each of those 4 years. A study of student evaluations for these professors and courses was also revealing. To illustrate this, consider the following student comments. Each comment is from a different year. The indicated percentage shows those who went on to complete a mathematics degree.

- "Dr….. taught well and always helped outside of class. However, his grading was not as helpful. He often did not explain his marks on the homework." *46.6% success rate*
- "'Don't you get this?' and 'Isn't this obvious?' and other similar questions do a good job of building walls between you and students. These also take away bits of confidence from us each time. Not only do you easily follow rabbit trails, but you often create your own out of a misunderstood question from a student." *50% success rate.*
- "The instructor demonstrated a thorough understanding of the material and methods. However, he did not effectively relate this knowledge to the class in a way that was conducive to student progress. I had a very hard time learning from him. The class seemed to be a lot of magic and hand waiving [sic]." *52.2% success rate.*
- "I think the group concept worked very well. It turned out to be far more effective than if he had lectured to us. We discovered things on our own, and when we had questions, he was always extremely helpful." *68.7% success rate.*

While some level of failure is inevitable for students who simply do not have the passion and ability for proofs, remarkable improvements may be possible with the right teaching style. Encouraging students to work collaboratively and allowing them to redo and continuously improve on their work is critical to the success of this course and our programs.

Assessment via exit interviews was useful not only for what it revealed about our Analysis sequence. The following points show how assessment can be both rewarding and a refutation of commonly held misconceptions.

- Roughly 90% of graduating students reported that their professors were in general very helpful.
- Approximately 95% of students don't feel advising is an issue for them. The program is self-explanatory and the only times they had complaints were when extremely technical advising questions could not be answered quickly and definitively. This would be very hard to remedy in any systematic way. This is in contrast to administration's claims that the campus is doing a poor job with advising.
- Student tastes are very mixed. In addition to being the most difficult course, Analysis occurred over 75% of the

time as either one of the most relevant or one of the least relevant courses taken.

- Teamwork is very valuable for learning and the formation of study groups is very common. However, in some courses where work is graded this way there is a sentiment that not all team members share the workload. Consequently instructors may wish to promote group study habits without grading in groups.

- No one was able to say that the department teaches the use of technology in any systematic way.

- In contrast to the big push in mathematics for new forms of pedagogy, successful students overwhelmingly prefer the standard lecture format over other more interactive classroom formats.

- Students would like more information about career opportunities and internships as they proceed through the program. Mathematics students (as opposed to those in statistics) are often very uninformed about possible career paths. In the short term, the possibility of some graduate courses and the opportunity for part-time teaching is often extremely appealing.

Use of findings

From our key finding about Analysis I & II, it became clear that the department must carefully try to understand what is working in these courses and what needs to be improved. To this end the following conclusions were drawn:

- Analysis cannot be taught like other courses in the sense of lecture, homework, lecture, homework, exam. Instead, students must be given practice in writing proofs with constant feedback and many opportunities to redo and discuss their work. Only those instructors regarded as excellent one-on-one mentors who have lots of time outside of class should teach these courses. Group studying should be encouraged.

- Students who are not prepared for Analysis should be encouraged to a more gentle introduction to proofs course like the three semester secondary education pre-service sequence. Recent discussions have proposed the creation a new transition to proofs course specifically for mathematics and statistics majors.

- Students must have more frequent access to Analysis. At the beginning of this assessment round Analysis I was offered every fall and Analysis II every spring. Once a year is just not enough for students who need to repeat these courses. The department has now gone to a system of offering each class every semester. This means that when an instructor is a poor fit with a class or a student is simply a slow learner, he or she can get right back in

the game the next semester with a different professor. Conversely, when a student likes his or her professor in Analysis I they can follow them to Analysis II.

Truth in Advertising. Large enrollment increases in Analysis without a corresponding rise in graduation rates told us that we had a retention problem. On reflection, we realized that some students were attracted to mathematics because of their love for computations. Naturally, such students would be disillusioned when they encounter the shock of Analysis and it may not be appropriate to continue. We should not just give up on them but we also cannot expect to win every battle. This led us to rethink the course descriptions for these courses. We illustrate this with Analysis I :

Before Assessment: (Analysis I) A re-examination of the calculus of functions of one-variable: real numbers, convergence, continuity, differentiation and integration. Prerequisite: Calculus III.

To the unaware this could have been interpreted as a kind of calculus IV taught in the usual way. Our new description will hopefully remind these weaker students of those scary passages in their calculus book which they always tried to avoid.

After Assessment: (Analysis I) An examination of the theory of calculus of functions of one-variable with emphasis on rigorously proving theorems about real numbers, convergence, continuity, differentiation and integration. Prerequisite: Calculus III.

Reflections, lessons and next steps

In hopes that other departments can effectively benefit from our experiences, we present some lessons learned the hard way. Key points are highlighted.

Mathematics departments have a diversity of service missions unparalleled in other disciplines. This unique role can have a critical impact on departments when they are faced with assessment mandates. Dealing with assessment in a meaningful way that is not intrusive on the important professional practices of a large faculty, is an important consideration. *Mathematics departments should monitor the interpretation and formulation of assessment mandates on their campuses very carefully.* Some universities will be happy with any assessment while others will demand a specific kind or even mandate that departments assess everything in their service role. In contrast, most accreditors will not be too particular about what assessment departments do and this will mean that there are many opportunities to appeal to the expertise of specific faculty without engaging the entire department. For example, in our case experts in mathematics education have NSF funding to consider the assessment and placement of students in lower division pre-calculus courses. Had we been proactive with the administration and their consultations with the accrediting agencies, we could have made this the centerpiece of our assessment efforts. The administration's lack of support of these placement efforts could be disastrous in the long run.

The beginning of the assessment procedure is deceptively simple and inviting. A mission statement and broad description of learning objectives are easy to agree on. However, *faculty involved in assessment should take care in distinguishing between a broad mission statement and extensive learning outcomes that shape what is expected of students and what faculty strive for and those aspects of a curriculum that can be reasonably evaluated.* The former is great to advertise your program, but for the latter it is essential to keep things manageable. For example, in our case the learning outcomes which require an appreciation for the interconnections of various areas of mathematics or the interplay between pure and applied mathematics are certainly important. However, they are impossibly broad to assess in a meaningful way.

Grade tracking is a useful alternative to a full blown portfolio system. There are, however, many ways this can be done, but care must be taken to factor in the grading standards different instructors have. In hindsight, for example, we could see that an expanded system where sequencing and instructor information is also used could have been very helpful.

We have learned that *there are several important techniques to making effective use of exit interviews.* At UNR we give students the exit interview questions in advance and they are asked to spend some time thinking about their responses before the interview. In addition, to get them to further reflect it is often helpful to have the interviews conducted in groups. This allows them to ponder each other's responses and further elaborate or provide counterpoint to the discussions. However, *exit interview questions and techniques should be continuously refined to be aligned with department goals.* For example, in the future we might well try to align our questions more directly with out learning outcomes (e.g., "Did your experience provide you with an appreciation of the interconnections of various mathematical disciplines?")

As we have explained, Analysis represented a serious problem for our majors. *When learning to prove things, hands-on activities with group studying and the opportunity to try again and again are essential.*

In retrospect, assessment has not been hard because our deficiencies and the key nature of Analysis sequence were

obvious. However, by appealing to the faculty's desire to improve rather than to an attempt to motivate them to assess, we have simultaneously harnessed faculty energy for change and kept our administration satisfied.

We now need to document that these improvements are effective. Given the long range approach which we used to analyze the problem it will take some time (e.g., 4-5 years) to see the full effect of these modifications. In the meantime, we should begin anew with some other problematic aspect of our curriculum. We close with some brief musings on this.

- UNR has an office of institutional analysis that can access data for UNR and K–12. For instance, the ability to track student performance in our major programs in relation to various high school indicators (such as GPA or the highest math taken or whether math was taken during the senior year), could be exceedingly revealing. We should use this office to give us ideas for future assessment.

- We now have an additional mandate to assess all of our graduate degree programs. We will rely heavily on exit interviews and follow up alumni surveys. The additional perspective gained several years after graduation should be useful and the lessons learned in our exit interviews of undergrads will be valuable.

- While the department's major is not highly populated, we do service a very large number of mathematics minors. Could they be evaluated? What would be the objectives in this case? How can we know if we are helping students achieve these goals or if their own major programs deserve the credit? Since all the courses taken by our minors are also taken by our majors, such an investigation would also provide valuable insight to our major programs from a different perspective.

The secondary information gained through our exit interviews is also a great source of ideas for future assessment. In this regard the department could consider several approaches:

- Development of a survey for those who drop out of our major programs and courses.

- Alumni interviews might help shed light on the following: Should programs in applied mathematics and statistics be de-emphasizing the role of proofs in favor of other preparations? How much technology should we be teaching?

Acknowledgements. While clearly this account is a story of a chair's work to get his faculty to assess themselves, it was not a task done in isolation. I owe a huge thanks to many people. Faculty members Bruce Blackadar, Mark

Meerschaert, Jeff Mortensen, Chris Herald, Tom Quint, and Tomasz Kozubowski provided careful consideration of my frequent musings and proposals on possible approaches to assessment. Jerry Johnson's willingness to attend several SAUM workshops and share his thoughtful insights and impressive background in numerical literacy and NSF based pre-calculus assessment was invaluable. It was Jerry who solved the mystery of our increased enrollments in Analysis by non-majors. Finally, I owe a big debt to the Director of University Assessment Dr. John Mahaffy who, despite my frequent complaints and no doubt exasperating noncompliance, stayed the course and continued to encourage and prod me to success.

Appendix A. Learning Outcomes

In the process of obtaining the broad base of content and skill based knowledge described in the department's mission statement, the student will also achieve a significant number of the following skills (whose precise emphasis will again vary by program option)

1. *Problem Solving Skills:* The ability to make precise sense of complicated situations in a variety of subject areas. This can include situations where there is too much or too little information and solutions will involve a variety of techniques from a range of different subjects.

2. *The appropriate use of technology*: This involves a range of activities from making routine calculations to modeling real world phenomena to experimentation with mathematical systems for the purpose of formulating conjectures and producing counterexamples.

3. *Modeling:* translating the real world into mathematical models that can be explored with technology and theoretical considerations. The results of such investigations must then be communicated to the lay mathematician or lay professional in a concise and effective manner.

4. *Methods of Proof:* Learning how to make rigorous mathematical arguments including how to both prove and disprove conjectures. This will also include reading mathematics and checking the proofs of others for completeness and correctness.

5. *Statistical Analysis:* This includes a firm understanding of a broad range of issues from the design of experiments to hypothesis testing and prediction to an understanding of when circumstances require consultation with more experienced statisticians.

6. *Working with axiomatic systems:* Proving basic facts from the axioms and determining if given examples satisfy the axioms. (Examples include but are not limited to the axioms for a vector space, groups & rings, a topological space, or those of Euclidean and non-Euclidean geometries.)

7. *Equivalence Relations and Equivalence Classes:* Understanding when operations are well defined on these classes and how functions either are or are not well defined with other structures.

8. *Appreciation for the interconnections of various mathematical disciplines:* This will include but is by no means limited to exposure to problems whose solutions involve a variety of disciplines as well as seeing techniques and modes of thought common to many subjects.

9. *Appreciation for the connections between applied and pure mathematics:* Understanding why distinctions between the two areas are not precise and how applied questions often generate large amounts of theoretical research.

10. *Appreciation for the career and educational opportunities for mathematics and statistics majors:* The realization that many professions value the problem solving skills of mathematicians and their ability to quickly learn and adapt to new situations. Likewise, statisticians possess a unique ability to interpret and gather highly useful and intricate quantitative descriptions of a vast set of circumstances. Both of these could certainly include internship experiences and a discussion of possibilities for advanced degrees.

Appendix B. Undergraduate Exit Interview Questions

Dear graduating senior,

Congratulations on your imminent graduation. Your hard work is about to pay off!

In order to improve our programs and curricula, we would like you to consider the following questions. When you meet with the chair soon he would like to discuss your feelings about each of these points. Your thoughtful and honest responses will be most appreciated. If possible, we may try to conduct these interviews along with other students. We have found that their responses often trigger deeper consideration and explanation of your sentiments as well.

Thanks you so much for your time and perspective on this very important matter.

1. What is your program option?
2. Are you getting a second degree?
3. What is your minor and how do you think these courses helped or hurt you?
4. What are your future career or educational plans?
5. What course(s) did you find most useful for your education?
6. What course(s) did you find least relevant to your education?
7. What course(s) did you find the most challenging?
8. Did you have instructors that you found to be most effective? Least Effective? Why?
9. What teaching styles did you think were most effective? Least effective? Why?
   - Traditional Lecture.
   - Computer Demonstrations.
   - Lab experiences/assignments.
   - Group Work.
   - Use of email.
   - Use of the internet.
   - Other.
10. How important was learning to write and read correct mathematical proofs?
11. Did you learn to use technology effectively and how important was this to your education?
12. How effective was the advising you received?
13. How important was the core curriculum to your education?
14. Do you have any general complaints or compliments about your experiences?

Assessing the Mathematics Major with a Bottom-Up Approach: First Step

Sarah V. Cook
Department of Mathematics and Statistics
Washburn University
Topeka, KS
sarah.cook@washburn.edu

Abstract. This paper looks at the developing stage of an assessment plan at a small liberal arts university. This plan assesses the calculus sequence through the use of pre and post-tests. In this paper, we discuss the selection of the questions used for the exams, the implementation of the assessment tool, and results obtained.

What did we hope to accomplish?

Washburn University is a municipally supported, state assisted university located in Kansas's capital city of Topeka. The university is comprised of six major academic units; the College of Arts and Sciences, the School of Law, the School of Business, the School of Nursing, the School of Applied Studies and the Division of Continuing Education and has an undergraduate population of around 7000 students. The number of mathematics majors is small, with approximately eight majors graduating per year. These majors are spread out over three tracks: pure mathematics, secondary education, and actuarial science.

Although the three tracks have a common set of courses, there is enough variance amongst the degrees to make assessing the major quite a challenge. With such a small number of overall majors, there are not enough students in any one track to justify implementing three completely different assessment plans. Our challenge has been to find an assessment plan that assesses the core knowledge all majors have, yet also addresses some of the specific skills inherent to the individual tracks.

The department decided it would be easiest to start assessing our majors from the bottom, where courses for the three tracks are the same. With this in mind, we decided to focus our initial assessment activities on the calculus sequence.

What did we do?

In March 2003 two members of our department attended the first part of the SAUM Workshop #3 in Phoenix where they developed a pre and post-test assessment system for the calculus sequence. The hope was that a pre and post-test system would provide information on student learning while also giving valuable feedback on student retention of knowledge. Ideally, pre-test questions would test specific prerequisite skills that would be needed for completing post-test problems for the same course. Further, some post-test questions from Calculus I and II would be used as pre-test questions in Calculus II and III respectively. Thus, the pre-test would not only give us a basis of comparison for knowledge gained in the current course, but also provide us with information on retention of knowledge from the prior course.

The pre-tests were implemented in the form of a review quiz given within the first week of the calculus course. This review quiz was counted in the student's overall quiz score for the semester. It should be noted that the instructors who teach the calculus sequence were already spending part of the first week reviewing prerequisite skills and giving some type of review assignment. Hence, the implementation of a review quiz fell in quite naturally with the instructors' plans. Post-test questions were simply embedded in the final examination.

After conferring with several department members, it was decided that pre-tests would consist of 6–8 questions. Most, if not all, of these prerequisite skills would be required to solve a post-test problem. Additionally post-tests for Calculus I and II would contain some problems from the Calculus II and III pre-tests. Pre and post-test questions were scored on a scale of 0–4 as follows:

0 – completely wrong/incorrect approach taken

1 – some correct work, but mostly incorrect

2 – general idea with some mistakes

3 – correct approach with minor error(s)

4 – completely correct

One member of our department wrote 10–12 pre-test questions for each of the calculus courses. These questions were then shared with other faculty members. From this initial set of problems, the department agreed on eight questions for each of the pre-tests in Calculus I and II and seven questions for Calculus III. These pre-test questions were administered in the Fall 2003 semester. For all pre-test quizzes and post-test questions see the SAUM website.)

To illustrate the pre/post-test system, let us begin with Calculus I. Pre-test questions were written to test specific algebraic and trigonometric skills that would be needed to solve standard calculus problems, which were then administered on the post-test. As examples, consider the following questions given on the Calculus I pre-test:

- Find the equation of the line through the point $(5, -7)$ and having slope $m = 11/3$.
- Rationalize the denominator: $x - 6/(\sqrt{x+3} - 3)$.
- Solve the equation for all x in $[0, 2\pi]$: $1 - 2\cos x = 0$.

The corresponding post-test questions are as follows:

- Find an equation of the line tangent to $f(x) = x^2 + 3x - 13$ at $x = 2$.
- Find the limit, if it exists. $\lim_{x \to 2} x - 6/(\sqrt{x+3} - 3)$.
- Find the critical numbers of the function $h(x) = x - 2\sin x$ on the interval $[0, 2\pi]$.

As demonstrated with these three examples, the algebraic and trigonometric skills necessary to correctly answer the post-test calculus problems depends on the students' ability to correctly solve the pre-test problems. Similar to these examples, all pre/post-test problems for Calculus I directly connect calculus knowledge to prerequisite skills.

Our initial attempt at obtaining such a clear connection of pre and post-test questions in Calculus II and III was not as clean as in Calculus I. There are several reasons for this.

First, our attempt at using pre-tests questions in Calculus II and III for the dual purpose of testing prerequisite skills needed to solve post-test problems and of testing retention of prerequisite knowledge caused instructors to feel limited when writing final examinations. For example, the Calculus I final contained eight post-test questions, which were directly connected to that course's pre-test problems. Separate from these problems, the final examination also had four questions that appeared on the Calculus II pre-test. Thus twelve problems on the Calculus I final exam were some form of an assessment question which left the instructor very little freedom in writing the exam.

Second, only the pre-tests were written before the semester began with the understanding that instructors of the calculus courses would write post-test questions which used the same skills as the prerequisite problems. The instructor who wrote the review quizzes is the same instructor who taught Calculus I in Fall 2003. This instructor planned the types of problems that would be on the Calculus I final examination before writing the pre-tests and then geared Calculus I pre-test questions toward the algebraic and trigonometric skills needed to solve these final examination problems. When writing the pre-test questions for Calculus II and III, the instructor tried to anticipate final exam problems and wrote pre-test questions with these skills in mind. However, these final exam problems were not necessarily the same types of questions the faculty who taught these courses wanted to use. For example, a Calculus II pre-test question was to evaluate the integral

$$\int_2^3 \frac{dx}{(7 - 4x)^2}$$

The instructor who wrote the pre-test anticipated that the Calculus II final would include a question which asked the student to determine if the integral

$$\int_0^2 \frac{dx}{(7 - 4x)^2}$$

converged. Due to the length of the final examination, the Calculus II instructor opted not to use such a problem on the final.

The largest problem we had in mimicking pre-test questions on a post-test was in Calculus II. Core topics in Calculus II such as integration techniques, series convergence tests and polar function graphing are difficult to compare with a pre-test question. The only Calculus I prerequisite for these types of problems is basic integration skills. While there are several algebraic and trigonometric skills necessary to solve these problems, it is common for students to learn these skills at the same time they learn the calculus. Because of this, the department did not feel we could adequately compare these types of standard Calculus II problems to a pre-test question.

The problems mentioned above have led us to make the following changes to the pre/post-test structure. The Calculus I assessment will remain largely as is, except that the post-test will mimic at least four, and not necessarily all eight, of the pre-test problems. These four questions will be up to the instructor's choosing and may rotate on a yearly basis. The post-test will also include four questions that appear on the Calculus II pre-test.

Calculus II will no longer be directly assessed with its own pre and post-test. Instead, the Calculus II pre-test will be used as a retention indicator for Calculus I. Three to four problems on the Calculus II post-test will be identical to problems on the Calculus III pre-test to test retention of Calculus II knowledge.

In addition to serving as a retention indicator for Calculus II, the Calculus III pre-test will contain four problems on prerequisite skills needed to correctly solve Calculus III post-test questions. As with Calculus I, these four problems may vary depending on the instructor.

What did we learn?

With only limited data and the problems we have faced in the first administering of the exams, it is difficult to comment on the effectiveness of our assessment tool. It is our hope that through time the pre and post-test system will give us a clearer understanding of whether or not our students are retaining essential skills from one course to the next and whether or not students are able to expand upon prerequisite information and combine it with newly acquired skills to solve a new set of problems.

Data regarding the means of student responses to pre and post-test questions can be found in Appendix A. Since the department feels that the best pre/post-test system occurred in Calculus I, comments on insights will refer to this course.

At this time it is appropriate to mention that the grading of post-test questions was done in two methods. One score was given for how the student performed overall on the problem, considering both the calculus and algebraic/trigonometric skills required. Another score was given on how the student performed based solely on the algebra and trigonometry and disregarding the calculus portion.

Of the students in Fall 2003 who took both the pre and post test ($n = 37$), the mean scores considering only algebraic and trigonometric skills improved on six of the eight problems and fell slightly on the other two. When comparing pre-test algebraic and trigonometric means to post-test means on calculus skills, there was improvement in only four of the eight questions.

For students who passed the course with a D or better ($n = 32$), the mean scores considering only algebraic and trigonometric skills improved on seven of the eight pre-test questions. The post-test means of the calculus skills still showed improvement on only four of the eight questions.

Only 17 students who took the Calculus I pre and post-test continued with Calculus II in the Spring 2004 semester. The Calculus II pre-test contained seven questions that tested similar concepts on the Calculus I final exam. Of these seven questions, the means of the calculus skills improved on four questions. The means on the remaining three questions dropped from 3.24, 3.00, and 3.71 to 2.71, 1.82, and 1.59 respectively. It was somewhat alarming to see such a drop in nearly half the questions. However, these were the last three questions on the review quiz. The students were allotted 20 minutes to complete the quiz and many students did not finish the last problems.

At this time, the department sees no need to make changes in the way the calculus sequence is taught. However, we do plan to continue to use the pre/post-test structure to track student learning and retention. If we notice problematic trends, we will address those problems at that time.

Along with the changes already mentioned in the previous section, we intend to more closely involve all calculus instructors in the selection of both pre and post-test questions. Also, we will allow ourselves flexibility in that post-test questions need not reduce to the identical problems on the pre-test. It is enough to have problems similar in nature that test the same skills. Finally, a logistical change is that the post-test questions will be the beginning questions on the final exams and given in the same order as the pre-test question they are associated with. This will aid in the grading and recording of scores.

Appendix A. Pre- and Post- Assessment Results for Calculus

Calculus I

The table below gives the results of students who took both the Calculus I pre- and post-test in Fall 2003 ($n = 37$).

Problem	Pre-Test Mean	Post-Test Mean Alg/Trig skills	Post-Test Mean Calculus skills
Line/tangent	3.65	3.32	2.89
Inequality/increasing	2.76	3.51	3.43
Equation/extrema (rational exponents)	2.73	3.32	2.49
Rationalize/limit	2.38	3.49	3.46
Complex fraction/limit	1.86	2.59	2.49
Simplify/integrate (divide)	1.92	3.11	2.49
Trig Equation/critical numbers	2.54	2.84	2.16
Ladder/ Rate of Change	3.76	3.62	3.11

The following table give the results of students who took both the Calculus I pre- and post-test in Fall 2003 and passed Calculus I in Fall 2003 with a grade of D or better ($n = 32$).

Problem	Pre-Test Mean	Post-Test Mean Alg/Trig skills	Post-Test Mean Calculus skills
Line/tangent	3.69	3.50	3.09
Inequality/increasing	2.91	3.78	3.69
Equation/extrema (rational exponents)	2.84	3.50	2.78
Rationalize/limit	2.22	3.91	3.88
Complex fraction/limit	2.06	2.84	2.72
Simplify/integrate (divide)	2.13	3.41	2.75
Trig Equation/critical numbers	2.56	2.94	2.44
Ladder/ Rate of Change	3.72	3.84	3.25

Calculus II

The table below gives the results of students who took both the Calculus II pre- and post-test in Fall 2003 ($n = 12$). All of these students passed the course.

Problem	Pre-Test Mean	Post-Test Mean Alg/Trig skills	Post-Test Mean Calculus skills
Product/chain rule	3.67	3.58	3.67
Nested chain rules	3	—	—
Definite integral/u-sub	2.33	—	—
Critical numbers (Trig)	2.17	1.42	2.5
Intersection points	4.00	4.00	4.00
Integrate after dividing	1.58	—	—
Inverse trig integration	1.92	3.17	2.67
Ladder/ Rate of Change	1.00	—	—

Calculus III

The table below gives the results of students who took both the Calculus III pre- and post-test in Fall 2003 ($n = 12$). All of these students passed the course.

Problem	Pre-Test Mean	Post-Test Mean Calc II skills	Post-Test Mean Calc III skills
Integration by parts	3.75	3.58	3.08
u-sub integration	3.42	—	—
Partial Fractions	2.25	—	—
Area between graphs	3.08	3.67	2.67
L'Hopitals	3.17	—	—
Equation of a plane	2.25	3.17	3.08
Parametric Equations	2.50	—	—

Combined Results (Fall 2003 & Spring 2004)

The following table gives mean scores on the Calculus I post-test and corresponding questions on the Calculus II pre-test for those who took Calculus I in Fall 2003 and are taking Calculus II in Spring 2004 (n=17).

Problem on Calculus II pre-test	Calc I post-test mean	Calc II pre-test mean
Product/chain rule	2.82	3.82
Nested chain rules	3.12	3.65
Definite integral/u-sub	3.29	3.11
Critical numbers (Trig)	2.41	3.18
Intersection points		3.76
Integrate after dividing	3.24	2.71
Inverse trig integration	3.00	1.82
Ladder/ Rate of Change	3.71	1.59

The following table gives mean scores on the Calculus II post-test and corresponding questions on the Calculus III pre-test for those who took Calculus II in Fall 2003 and are taking Calculus III in Spring 2004 ($n = 6$). (*Note*: There was an error made in Spring 2004 and the wrong version of the Calculus III pre-test was given. Because of this, the questions regarding L'Hopital's rule and graphing a curve from its parametric equations were not given on the pre-test.)

Problem on Calculus III pre-test	Calc II post-test mean	Calc III pre-test mean
Integration by parts	3.00	2.83
u-sub integration	2.83	0.83
Partial Fractions	2.83	2.50
Area between graphs	4.00	3.00
L'Hopital's Rule	2.83	
Equation of a plane	2.50	1.67
Parametric Equations	3.67	

Personnel

SAUM Assessment Workshops

William E. Haver
Department of Mathematics
Virginia Commonwealth University
Richmond, VA
whaver@mail2.vcu.edu

Four multi-session assessment workshops were conducted by SAUM in conjunction with the MAA PREP Program. Each workshop comprised two or three sessions for a total of four to six days. Colleges and universities were invited to nominate teams to participate in the workshop series. The workshop provided support to the members of the teams to help them lead their departments in the conceptualization, formulation, and implementation of assessment activities for some component of their academic programs.

Typically, prior to the first session the teams prepared a short description of the assessment program that was either planned or in progress at the team's home department. During the first session participants considered many aspects of assessment and heard from others with extensive experience in assessment, including Bernard Madison, Lynn Steen, and Peter Ewell, as well as individuals who served as mentors at various workshops: Bonnie Gold, Sandra Keith, Bill Marion, Bill Martin, Kathy Stafford-Ramus, and Rick Vaughn. A number of these individuals were participants in one workshop series and then served as mentors in subsequent workshop series.

Workshop participants formalized the next steps in the assessment process and agreed on benchmarks to meet before the subsequent workshop session. The participating teams also made commitments to add their experiences to the developing literature. Indeed a large number of the SAUM Case Studies reported in this volume came from work that was initiated during the SAUM/PREP workshops.

As a part of the formative evaluation of the workshop series, Peter Ewell reported on an open discussion held at the conclusion of a joint session of two of the workshop series. He stated that "participants who spoke were overwhelmingly satisfied with the experience, citing that it had enabled them to make real progress in building an assessment project at the departmental level on their own campuses and find colleagues elsewhere who could support them."

As Ewell noted, participants also reported that:

- Multiple workshops meant that participants could learn things, go back to their campuses and apply them, then return to demonstrate and share. Virtually everyone who provided comments on the multiple-workshop design maintained that a format featuring several encounters over time provided a far better learning experience than a "one-shot" workshop.

- The need to present campus progress at each workshop provided peer pressure to keep the process moving. Teams knew that they were going to have to present something publicly, so worked hard to have progress to report.

- Working with other campuses helped build a feeling of being part of a larger "movement" that had momentum.
- The workshop helped legitimize the work of developing assessment at participating campuses when the team got to work back home. The sense of being part of a larger, recognized, NSF-funded project was important in convincing others that this effort was important.

A prerequisite for participation was that teams agreed to share what they learned from others. Members of workshop teams listed below can be considered good sources of information and consultation for others interested in learning more about assessment.

Workshop Number 1

- January 10–11, 2002; San Diego, Calif.
- May 22–25, 2002; Richmond, Virginia
- January 19–20, 2003; Towson, Maryland

Allegheny College
Team: Ronald Harrell, Tamara Lakins

Canisius College
Team: Chris Kinsey, Dietrich Kuhlmann

Cloud County Community College
Team: Tim Warkentin, Mark Whisler

Houston Community College
Team: Jacqueline Giles, Cheryl Peters

Jacksonville University
Team: Robert Hollister, Marcelle Bessman

Mount Mary College
Team: Abdel Naser Al-Hasan, Patricia Jaberg

North Dakota State University
Team: Jorge Calvo, Dogan Comez, William Martin

Paradise Valley Community College
Team: *Rick Vaughn, Larry Burgess*

Portland State University
Team: Paul Latiolais, Karen Marrongele, Jeanette Palmiter

Saint Peter's College
Team: Brian Hopkins, Eileen Poiani, Katherine Stafford-Ramus

United States Military Academy
Team: Alex Heidenberg, Mike Huber, Joseph Meyers, Kathleen Snook, Frank Wattenberg

University of Arizona
Team: Pallavi Jayawant, David Lomen, Peter Wiles

University of Arkansas at Little Rock
Team: James Fulmer, Thomas McMillan

University of Central Oklahoma
Team: David Bolivar, Carol Lucas

University of Texas System
Team: D. L. Hawkins, Jerry Mogilskj, Betty Travis

Virginia Commonwealth University
Team: Aimee Ellington, Reuben Farley, Kim Jones

Workshop Number 2

- July 29–31, 2002 Burlington, Vermont
- January 19–20, 2003 Towson, Maryland
- January 7, 2004, Phoenix, Arizona

Arapahoe Community College
Team: Jeff Berg, Erica Johnson

Columbia College
Team: Nieves McNulty

Keene State College
Team: Vince Ferlini, Dick Jardine

Long Island University, Southampton
Team: Russell Myers

Mercer University
Team: Tony Weathers

Mitchell College
Team: Ann Keating

North Carolina State University
Team: Jeff Scroggs

Pomona College
Team: Everett Bull, Shahriar Sharhari

South Dakota State University
Team: Dan Kemp, Ken Yokom

University of Nevada, Reno
Team: Jerry Johnson, Ed Keppelman

Workshop Number 3

- March 14–16, 2003, Phoenix, Arizona
- January 5–7, 2004, Phoenix, Arizona

Arizona Western University
Team: Leroy Cavanaugh, Brian Karasek, Daniel Russow

Colorado Schools of Mines
Team: Barbara Moskal, Alyn Roskwood

Eastern New Mexico University
Team: Regina Aragon, Tom Brown, John George

James Madison University
Team: David Carothers, Robert Hanson

Kirkwood Community College
Team: David Keller, John Weglarz

Point Loma Nazarene University
Team: Jesus Jimenez, Maria Zack

Regis University
*Team:*Richard Blumenthal, Carmen Johansen, Mohamed Lofty, Jennifer Mauldin

Rhode Island College
Team: James Schaefer

Rhodes State College
Team: Judy Giffin, Mary Ann Hovis

San Jose State University
Team: Trisha Bergthold, Ho Kuen Ng

University of Arkansas at Little Rock
Team: Melissa Hardeman, Tracy Watson

University of Wisconsin Oshkosh
Team: John Koker, Jennifer Szydlik., Steve Szydlik

Washburn University
*Team:*Sarah Cook, Donna LaLonde

West Virginia University
Team: Eddie Fuller, James Miller

Workshop Number 4

- March 5–8, 2004, High Point, NC
- January 8–10, 2005, Atlanta, GA
- January 10–11, 2006, San Antonio, TX

American University
Team: Virginia Stallings, Matt Pascal, Dan Kalman

Belmont University
Team: Mary Goodloe, Barbara Ward, Mike Pinter

Centenary College of Louisiana
Team: Mark Schlatter, Derrick Head

Clayton College, State University
Team: Nathan Borchelt, Weihu Hong

College of St. Benedict/St. John's University
Team: Jennifer Galovich, Kristin Nairn, Gary Brown

College of St. Catherine
Team: Suzanne Molnar, Dan O'Loughlin, Adele Rothan

College of Wooster
Team: James Hartman, Pamela Pierce

Emporia State University
Team: Antonie Boerkoel, Connie Schrock, Larry Scott

Florida State University
Team: Monica Hurdal, Jerry Magnan

Gallaudet University
*Team:*fat Lam, Vicki Shank

Georgia College & State University
Team: Martha Allen, Hugh Sanders

Hood College
Team: Betty Mayfield, Doug Peterson, Kira Hamman

Kennesaw State University
*Team:*Virginia Watson, Mary Garner, Lewis VanBrackle

Lebanon Valley College
Team: Michael Fry, Christopher Brazfield, Patrick Brewer

Meredith College
Team: Timothy Hendrix, Jennifer Hontz, Cammey Cole

Metropolitan State College of Denver
Team: Larry Johnson, Ahahar Boneh

Oregon State University
Team: Thomas Dick, Barbara Edwards, Lea Murphy

Ouchita Baptist University
Team: Jeff Sykes, Steve Hennagin, Darin Buscher

Roosevelt University
Team: Jimmie Johnson, Ray Shepherd, Andy Carter

Southeastern Louisiana University
Team: David Gurney, Rebecca Muller

Southwest Missouri State University
Team: Yungchen Cheng, Shelby Kilmer,

Southwestern University
Team: Therese Shelton, John Chapman, Walt Potter, Barbara Owens

Stetson University
Team: Lisa Coulter, Hari Pulapaka

Texas A&M University–Texarkana
Team: Arthur Simonson, Dennis Kern

The College of New Jersey
Team: Edward Conjura, Cathy Liebars, Sharon Navard

United States Military Academy
Team: Alex Heidenberg, Michael Huber, Michael Phillips

Valparaiso University
Team: Kenneth Luther, Kimberly Pearson

Western Michigan University
Team: Tabitha Mingus, Jeff Strom, Dennis Pence

SAUM
Project Steering
Committee

Peter Ewell, project evaluator, is vice president of the National Center for Higher Education Management Systems (NCHEMS) in Boulder, Colorado. Ewell's work focuses on assessing institutional effectiveness and the outcomes of college; he has consulted with over 375 colleges and universities and twenty-four state systems of higher education on topics including assessment, program review, enrollment management, and student retention. Ewell has authored six books and numerous articles on the topic of improving undergraduate instruction through the assessment of student outcomes. In 1998 he led the design team for the National Survey of Student Engagement (NSSE) and currently chairs its Technical Advisory Panel. A graduate of Haverford College, Ewell received a PhD in political science from Yale University and subsequently served on the faculty of the University of Chicago.

Bonnie Gold, case studies editor, is professor of mathematics at Monmouth University, West Long Branch, New Jersey. Co-editor of *Assessment Practices in Undergraduate Mathematics* (MAA, 1999), Gold is currently editing a book on current issues in the philosophy of mathematics from the viewpoint of mathematicians and teachers of mathematics. Formerly Gold served as chair of mathematics departments at Wabash College (Indiana) and Monmouth University and has directed Project NExT (New EXperiences in Teaching) programs in both Indiana and New Jersey. Originator of MAA Online's Innovative Teaching Exchange, Gold has received an Open Faculty Fellowship from the Lilly Foundation and McLain-Turner-Arnold Award for Excellence in Teaching from Wabash College. A graduate of the University of Rochester, Gold received an MA from Princeton University and a PhD in mathematical logic from Cornell University.

William E. Haver, workshop coordinator, is professor of mathematics and former chair of the mathematics department at Virginia Commonwealth University in Richmond, Virginia. Haver is currently director of a Mathematics and Science Partnership program to prepare teachers to serve as full-time mathematics specialists/coaches in Virginia's elementary schools. The program has a large research component to help determine the impact of these teachers on student learning. Haver currently chairs the subcommittee on Curriculum Renewal Across the First Two Years (CRAFTY) of the Mathematical Association of America. Previously, he served as a senior program director at the National Science Foundation. A graduate of Bates College, Haver received a master's degree from Rutgers and a PhD in mathematics from SUNY-Binghamton.

Laurie Boyle Hopkins, case studies editor, is vice president for academic affairs and professor of mathematics at Columbia College in Columbia, South Carolina. Previously Hopkins served as chair of the department of mathematics and as chair of the faculty. In addition to assisting with SAUM workshops, Hopkins' work on assessment includes a presentation to the Southern Association of Schools and Colleges on using assessment to improve general education. Hopkins also participated in the Foundations of Excellence Project sponsored by the National Center for the First Year Experience and has been a leader in use of data from the National Survey of Student Engagement (NSSE). Hopkins holds a PhD degree in mathematics from the University of South Carolina and has pursued additional graduate work in computer science.

Dick Jardine, case studies editor, is chair of the mathematics department at Keene State College in Keene, New Hampshire. Jardine has been a member of the Mathematical Association of America's Committee on the Undergraduate Teaching of Mathematics (CUTM) and is currently on the Subcommittee on the Instructional Use of the History of Mathematics. He is co-editor of *From Calculus to Computers* (MAA, 2005), a compendium of 19th and 20th century sources and examples selected to help college instructors use the history of mathematics to enhance teaching and learning. In addition to serving as a SAUM workshop leader, Jardine was a keynote speaker at the State University of New York General Education Assessment Conference in April of 2005. Jardine earned his PhD in mathematics at Rensselaer Polytechnic Institute.

Bernard L. Madison, project director, is professor of mathematics at the University of Arkansas where he previously served as Chair of Mathematics and Dean of the J.W. Fulbright College of Arts and Sciences. During 1985-89, Madison directed the MS2000 project at the National Research Council, including the 1987 Calculus for a New Century symposium. Madison has served as Chief Reader for AP Calculus and as a member of the Commission on the Future of AP. Currently he heads the National Numeracy Network, leads workshops dealing with the mathematical education of teachers, and studies the articulation between school and college mathematics. Madison majored in mathematics and physics at Western Kentucky University and subsequently earned masters and doctoral degrees in mathematics from the University of Kentucky.

William A. Marion, case studies editor, is professor of mathematics and computer science at Valparaiso University in Valparaiso, Indiana, where he has taught for over twenty-five years. In 1991, Marion helped develop MAA's first policy statement on assessing undergraduate mathematics programs. Subsequently, he joined Bonnie Gold in editing *Assessment Practices in Undergraduate Mathematics* (MAA, 1999) and in directing several related mini-courses. Recently Marion co-chaired a national initiative on discrete mathematics for computer science students; conducted workshops for undergraduate mathematics faculty who teach such courses, and spoke on this work at the quadrennial World Conference on Computers in Education in South Africa. Marion holds a DA degree in mathematics from the University of Northern Colorado and has undertaken more than two years of graduate-level study in computer science.

Michael Pearson, principal investigator (2002–05), is associate executive director and director of programs and services at the Mathematical Association of America. His responsibilities include oversight of professional development activities for the Association, including SAUM. Prior to joining the Washington office of the MAA, Pearson served as associate head of the Department of Mathematics and Statistics at Mississippi State University. He also served in a variety of capacities in the Louisiana-Mississippi Section of the MAA. Beginning in the mid-1990's, Pearson became involved with the calculus reform move-

ment and, through several NSF-funded projects, explored various strategies for assessing student learning. In 1989 Pearson received his PhD in harmonic analysis at The University of Texas at Austin.

Thomas Rishel, principal investigator (2001–02), is professor of mathematics at Weill Cornell Medical College in Qatar. Previously, he served two years as associate director of the Mathematical Association of America, where he helped launch the SAUM project. Rishel came to MAA from Cornell University where he was a senior lecturer in mathematics and director of undergraduate teaching. He also as taught at Dalhousie University, Tokyo Kyoiku Daigaku, the University of Pittsburgh, and the University of Oregon. In addition to writing several papers on topology, Rishel is author of *Teaching First: A Guide for New Mathematicians* (MAA, 2000) and co-author of *Writing in the Teaching and Learning of Mathematics* (MAA, 1998). Rishel has a bachelor's degree from Youngstown State, and a master's and PhD from the University of Pittsburgh.

Lynn Arthur Steen is professor of mathematics and special assistant to the provost at St. Olaf College, in Northfield, Minnesota. Steen is the editor or author of many books on mathematics and education, including *Math and Bio 2010: Linking Undergraduate Disciplines* (2005), *Mathematics and Democracy* (2001), *On the Shoulders of Giants* (1991), and *Everybody Counts* (1989). His current work focuses on the transition from secondary to tertiary education, notably on the mathematical and quantitative requirements for contemporary work and responsible citizenship. Steen has served as executive director of the Mathematical Sciences Education Board (MSEB), as chairman of the Council of Scientific Society Presidents (CSSP), and as president of the Mathematical Association of America (MAA). Steen received his PhD in mathematics in 1965 from the Massachusetts Institute of Technology.

Appendix

CUPM Guidelines for Assessment of Student Learning

An updated and slightly revised reprint of "Assessment of Student Learning for Improving the Undergraduate Major in Mathematics," originally prepared by the Subcommittee on Assessment of the Committee on the Undergraduate Program in Mathematics (CUPM) of the Mathematical Association of America. Approved by CUPM on January 4, 1995, updated by CUPM on August 5, 2005.

Foreword to Revised Version

The original Guidelines, released in 1995, signaled an increased interest in assessment among the mathematics community. *Assessment Practices in Undergraduate Mathematics, MAA Notes #49,* was published in 1999 [9]. The MAA project Supporting Assessment in Undergraduate Mathematics (SAUM), supported by an NSF grant, began in 2001 (www.maa.org/saum/). PREP Workshops on assessment began in 2002. Contributed paper sessions on assessment, sponsored by SAUM, began in 2003. A second volume of case studies generated through the SAUM project was published by the MAA in 2005 [10].

These activities have led to a more mature understanding of assessment among the mathematical community as evidenced by its emphasis in *Undergraduate Programs and Courses in the Mathematical Sciences: CUPM Curriculum Guide 2004* [19]. We have revised the 1995 Guidelines and updated the references to reflect the changes in assessment practices and opportunities over the last ten years and to make it consistent with the language found in the *2004 CUPM Curriculum Guide.*

Ad hoc Committee to Review the Assessment Guidelines, 2005
    Janet Andersen, Hope College
    William Marion, Valparaiso University
    Daniel Maki, Indiana University

Preface to 1995 Original

Recently there has been a series of reports and recommendations about all aspects of the undergraduate mathematics program. In response, both curriculum and instruction are changing amidst increasing dialogue among faculty about what those changes should be. Many of the changes suggested are abrupt breaks with traditional practice; others are variations of what has gone on for many decades. Mathematics faculty need to determine the effectiveness of any change and institutionalize those that show the most promise for improving the quality of the program available to mathematics majors. In deciding which changes hold the greatest promise, student learning assessment provides invaluable information. That assessment can also help departments formulate responses for program review or other assessments mandated by external groups.

The Committee on the Undergraduate Program in Mathematics established the Subcommittee on Assessment in 1990. This document, approved by CUPM in January 1995, arises from requests from departments across the country struggling to find answers to the important new questions in undergraduate mathematics education. This report to the community is suggestive rather than prescriptive. It provides samples of various principles, goals, areas of assessment, and measurement methods and techniques. These samples are intended to seed thoughtful discussions and should not be considered as recommended for adoption in a particular program, certainly not in totality and not exclusively.

Departments anticipating program review or preparing to launch the assessment cycle described in this report should pay careful attention to the MAA *Guidelines for Programs and Departments in Undergraduate Mathematical Sciences* [1]. In particular, Section B.2 of that report and step 1 of the assessment cycle described in this document emphasize the need for departments to have:

a. A clearly defined statement of program mission; and

b. A delineation of the educational goals of the program.

The Committee on the Undergraduate Program in Mathematics urges departments to consider carefully the issues raised in this report. After all, our programs should have clear guidelines about what we expect students to learn and have a mechanism for us to know if in fact that learning is taking place.

— James R. C. Leitzel, Chair, CUPM, 1995

Membership of the Subcommittee on Assessment, 1995:

Larry A. Cammack, Central Missouri State University, Warrensburg, MO
James Donaldson, Howard University, Washington, DC
Barbara T. Faires, Westminster College, New Wilmington, PA
Henry Frandsen, University of Tennessee, Knoxville, TN
Robert T. Fray, Furman University, Greenville, SC
Rose C. Hamm, College of Charleston, Charleston, SC
Gloria C. Hewitt, University of Montana, Missoula, MT
Bernard L. Madison (Chair), University of Arkansas, Fayetteville, AR
William A. Marion, Jr., Valparaiso University, Valparaiso, IN
Michael Murphy, Southern College of Technology, Marietta, GA
Charles F. Peltier, St. Marys College, South Bend, IN
James W. Stepp, University of Houston, Houston, TX
Richard D. West, United States Military Academy, West Point, NY

I. Introduction

The most important indicators of effectiveness of mathematics degree programs are what students learn and how well they are able to use that learning. To gauge these indicators, assessment — the process of gathering and interpreting information about student learning — must be implemented. This report seeks to engage faculty directly in the use of assessment of student learning, with the goal of improving undergraduate mathematics programs.

Assessment determines whether what students have learned in a degree program is in accord with program objectives. Mathematics departments must design and implement a cycle of assessment activity that answers the following three questions:

What should our students learn?

How well are they learning?

What should we change so that future students will learn more and understand it better?

Each step of an ongoing assessment cycle broadens the knowledge of the department in judging the effectiveness of its programs and in preparing mathematics majors. This knowledge can also be used for other purposes. For example, information gleaned from an assessment cycle can be used to respond to demands for greater accountability from state governments, accrediting agencies, and university administrations. It can also be the basis for creating a shared vision of educational goals in mathematics, thereby helping to justify requests for funds and other resources.

This report provides samples of various principles, goals, areas of assessment, and measurement methods and techniques. Many of the items in these lists are extracted from actual assessment documents at various institutions or from reports of professional organizations. These samples are intended to stimulate thoughtful discussion and should not be considered as recommended for adoption in a particular program, certainly not in totality and not exclusively. Local considerations should guide selection from these samples as well as from others not listed.

II. Guiding Principles

An essential prerequisite to constructing an assessment cycle is agreement on a set of basic principles that will guide the process, both operationally and ethically. These principles should anticipate possible problems as well as ensure sound and effective educational practices. Principles and standards from several sources (see references 2,3,4,5,and 6) were considered in the preparation of this document, yielding the following for consideration:

 a. Objectives should be realistically matched to institutional goals as well as to student backgrounds, abilities, aspirations, and professional needs.

 b. The major focus of assessment (by mathematics departments) should be the mathematics curriculum.

 c. Assessment should be an integral part of the academic program and of program review.

 d. Assessment should be used to improve teaching and learning for all students, not to filter students out of educational opportunities.

 e. Students and faculty should be involved in and informed about the assessment process, from the planning stages throughout implementation.

 f. Data should be collected for specific purposes determined in advance, and the results should be reported promptly.

III. The Assessment Cycle

Recommendation 1 in the *CUPM Curriculum Guide 2004* (1) states that mathematical sciences departments should

- Understand the strengths, weaknesses, career plans, fields of study, and aspirations of the students enrolled in mathematics courses

- Determine the extent to which the goals of courses and programs offered are aligned with the needs of students, as well as the extent to which these goals are achieved;

- Continually strengthen courses and programs to better align with student needs, and assess the effectiveness of such efforts

This recommendation leads to a culture of continual assessment within mathematics departments. Departments need to develop an assessment cycle that includes the following:

1. Articulate the learning goals of the mathematics curriculum and a set of objectives that should lead to the accomplishment of those goals.
2. Design strategies (e.g., curriculum and instructional methods) that will accomplish the objectives, taking into account student learning experiences and diverse learning styles, as well as research results on how students learn.
3. Determine the areas of student activities and accomplishments in which quality will be judged. Select assessment methods designed to measure student progress toward completion of objectives and goals.
4. Gather assessment data; summarize and interpret the results.
5. Use the results of the assessment to improve the mathematics major.

Steps 1 and 2 answer the first question in the introduction — what should the students learn? Steps 3 and 4, which answer the second question about how well they are learning, constitute the assessment. Step 5 answers the third question on what improvements are possible.

Step 1. Set the Learning Goals and Objectives

There are four factors to consider in setting the learning goals of the mathematics major: institutional mission, background of students and faculty, facilities, and degree program goals. Once these are well understood, then the goals and objectives of the major can be established. These goals and objectives of the major must be aligned with the institutional mission and general education goals and take into account the information obtained about students, faculty, and facilities.

Institutional Mission and Goals. The starting point for establishing goals and objectives is the mission statement of the institution. Appropriate learning requirements from a mission statement should be incorporated in the department's goals. For example, if graduates are expected to write with precision, clarity, and organization within their major, this objective will need to be incorporated in the majors' goals. Or, if students are expected to gain skills appropriate for jobs, then that must be a goal of the academic program for mathematics majors.

Information on Faculty, Students, and Facilities. Each institution is unique, so each mathematics department should reflect those special features of the institutional environment. Consequently, the nature of the faculty, students, courses, and facilities should be studied in order to understand special opportunities or constraints on the goals of the mathematics major. Questions to be considered include the following:

What are the expectations and special needs of our students?

Why and how do our students learn?

Why and how do the faculty teach?

What are the special talents of the faculty?

What facilities and materials are available?

Are mathematics majors representative of the general student population, and if not, why not?

Goals and Objectives of Mathematics Degree Program. A degree program in mathematics includes general education courses as well as courses in mathematics. General education goals should be articulated and well-understood before the goals and objectives of the mathematics curriculum are formulated. Of course, the general education goals and the mathematics learning goals must be consistent [6, pages 183–223]. Some examples of general education goals that will affect the goals of the degree program and what learning is assessed include the following:

Graduates are expected to speak and write with precision, clarity, and organization; to acquire basic scientific and technological literacy; and to be able to apply their knowledge.

Degree programs should prepare students for immediate employment, graduate schools, professional schools, or meaningful and enjoyable lives.

Degree programs should be designed for all students with an interest in the major subject and encourage women and minorities, support the study of science, build student self-esteem, ensure a common core of learning, and encourage life-long learning.

Deciding what students should know and be able to do as mathematics majors ideally is approached by setting the learning goals and then designing a curriculum that will achieve those goals. However, since most curricula are already structured and in place, assessment provides an opportunity to review curricula, discern the goals intended, and rethink them. Curricula and goals should be constructed or reviewed in light of recommendations contained in the CUPM Curriculum Guide 2004 [1].

Goal setting should move from general to specific, from program goals to course goals to assessment goals. Goals for student learning can be statements of knowledge students should gain, skills they should possess, attitudes they should develop, or requirements of careers for which they are preparing. The logical starting place for discerning goals for an existing curriculum is to examine course syllabi, final examinations, and other student work.

The CUPM Curriculum Guide 2004 includes learning goals such as:

Learn to apply precise, logical reasoning to problem solving Students should be able to perform complex tasks; explore subtlety; discern patterns, coherence, and significance; undertake intellectually demanding mathematical reasoning; and reason rigorously in mathematical arguments in order to solve complex problems.

Develop persistence and skill in exploration, conjecture, and generalization. Students should be able to undertake independent work, develop new ideas, and discover new mathematics. Students should be able to state problems carefully, articulate assumptions, and apply appropriate strategies. Students should possess personal motivation and enthusiasm for studying and applying mathematics; and attitudes of mind and analytical skills required for efficient use, appreciation, and understanding of mathematics.

Read and communicate mathematics with understanding and clarity.. Students should be able to read, write, and speak mathematically; read and understand technically-based materials; contribute effectively to group efforts; communicate mathematics clearly in ways appropriate to career goals; conduct research and make oral and written presentations on various topics; locate, analyze, synthesize, and evaluate information; create and document algorithms; think creatively at a level commensurate with career goals; and make effective use of the library. Students should possess skill in expository mathematical writing, have a disposition for questioning, and be aware of the ethical issues in mathematics.

Other possible learning goals include:

Nature of Mathematics. Students should possess an understanding of the breadth of the mathematical sciences and their deep interconnecting principles; substantial knowledge of a discipline that makes significant use of mathematics; understanding of interplay among applications, problem-solving, and theory; understanding and appreciation of connections between different areas of mathematics and with other disciplines; awareness of the abstract nature of theoretical mathematics and the ability to write proofs; awareness of historical and contemporary contexts in which mathematics is practiced; understanding of the fundamental dichotomy of mathematics as an object of study and a tool for application; and critical perspectives on inherent limitations of the discipline.

Mathematical Modeling. Students should be able to apply mathematics to a broad spectrum of complex problems and issues; formulate and solve problems; undertake some real-world mathematical modeling project; solve multi-step problems; recognize and express mathematical ideas imbedded in other contexts; use the computer for simulation and visualization of mathematical ideas and processes; and use the process by which mathematical and scientific facts and principles are applied to serve society.

Content Specific Goals. Students should understand theory and applications of calculus and the basic techniques of discrete mathematics and abstract algebra. Students should be able to write computer programs in a high level language using appropriate data structures (or to use appropriate software) to solve mathematical problems.

Topic or thematic threads through the curriculum are valuable in articulating measurable objectives for achieving goals. Threads also give the curriculum direction and unity, with courses having common purposes and reinforcing one another. Each course or activity can be assessed in relation to the progress achieved along the threads. Possible threads or themes are numerous and varied, even for the mathematics major. Examples include problem-solving, mathematical reasoning, communication, scientific computing, and mathematical modeling. The example of a learning goal and instructional strategy in the next section gives an idea of how the thread of mathematical reasoning could wind through the undergraduate curriculum.

Step 2. Design Strategies to Accomplish Objectives

Whether constructing a curriculum for predetermined learning goals or discerning goals from an existing curriculum, strategies for accomplishing each learning goal should be designed and identified in the curricular and co-curricular activities. Strategies should respect diverse learning styles while maintaining uniform expectations for all students.

Strategies should allow for measuring progress over time. For each goal, questions such as the following should be considered.

- Which parts of courses are specifically aimed at helping the student reach the goal?
- What student assignments help reach the goal?
- What should students do outside their courses to enable them to reach the goal?
- What should the faculty do to help the students reach the goal?
- What additional facilities are needed?
- What does learning research tell us?

The following example of a goal and strategy can be made more specific by referencing specific courses and activities in a degree program.

Learning goal. Students who have completed a mathematics major should be able to read and understand mathematical statements, make and test conjectures, and be able to construct and write proofs for mathematical assertions using a variety of methods, including direct and indirect deductive proofs, construction of counterexamples, and proofs by mathematical induction. Students should also be able to read arguments as complex as those found in the standard mathematical literature and judge their validity.

Strategy. Students in first year mathematics courses will encounter statements identified as theorems which have logical justifications provided by the instructors. Students will verify the need for some of the hypotheses by finding counterexamples for the alternative statements. Students will use the mathematical vocabulary found in their courses in writing about the mathematics they are learning. In the second and third years, students will learn the fundamental logic needed for deductive reasoning and will construct proofs of some elementary theorems using quantifiers, indirect and direct proofs, or mathematical induction as part of the standard homework and examination work in courses. Students will construct proofs for elementary statements, present them in both written and oral form, and have them critiqued by a mathematician. During the third and fourth years, students will formulate conjectures of their own, state them in clear mathematical form, find methods which will prove or disprove the conjectures, and present those arguments in both written and oral form to audiences of their peers and teachers. Students will make rational critiques of the mathematical work of others, including teachers and peers. Students will read some mathematical literature and be able to rewrite, expand upon, and explain the proofs.

Step 3. Determine Areas and Methods of Assessment

Learning goals and strategies should determine the areas of student accomplishments and departmental effectiveness that will be documented in the assessment cycle. These areas should be as broad as can be managed, and may include curriculum (core and major), instructional process, co-curricular activities, retention within major or within institution, and success after graduation. Other areas such as advising and campus environment may be areas in which data on student learning can be gathered.

Responsibility for each chosen area of assessment should be clearly assigned. For example, the mathematics faculty should have responsibility for assessing learning in the mathematics major, and the college may have responsibility for assessment in the general education curriculum.

Assessment methods should reflect the type of learning to be measured. For example, the Graduate Record Examination (GRE) may be appropriate for measuring preparation for graduate school. On the other hand, an attitude survey is an appropriate tool for measuring an aptitude for life-long learning. An objective paper-and-pencil examination may be selected for gauging specific content knowledge.

Eight types of assessment methods are listed below, with indications of how they can be used. Departments will typically use a combination of methods, selected in view of local program needs.

1. *Tests.* Tests can be objective or subjective, multiple-choice or free-response. They can be written or oral. They can be national and standardized, such as the GRE and Educational Testing Service Major Field Test, or they can be locally generated. Tests are most effective in measuring specific knowledge and its basic meaning and use.

2. *Surveys.* These can be written or they can be compiled through interviews. Groups that can be surveyed are students, faculty, employers, and alumni. Students can be surveyed in courses (about the courses), as they graduate (about the major), or as they change majors (about their reasons for changing).

3. *Evaluation reports.* These are reports in which an individual or group is evaluated. This may be done through a checklist of skills and abilities or may be a more holistic evaluation that includes descriptions of student performance. These can be completed by faculty members, peers, or employers of recent graduates. In some cases, self-evaluations may be used, but these tend to be of less value than more objective evaluations. Grades in courses are, of course, fundamental evaluation reports.

4. *Portfolios.* Portfolios are collections of student work, usually compiled for individual students under faculty supervision following a standard departmental protocol. The contents may be sorted into categories, e.g., fresh-

man or sophomore, and by type, such as homework, formal written papers, or examinations. The work collected in a student's portfolio should reflect the student's progress through the major. Examples of work for portfolios include homework, examination papers, writing samples, independent project reports, and background information on the student. In order to determine what should go in a portfolio, one should review what aspects of the curriculum were intended to contribute to the objectives and what work shows progress along the threads of the curriculum. Students may be given the option of choosing what samples of particular types of work are included in the portfolio.

5. *Essays.* Essays can reveal writing skills in mathematics as well as knowledge of the subject matter. For example, a student might write an essay on problem-solving techniques. Essays should contribute to learning. For example, students might be required to read four selected articles on mathematics and, following the models of faculty-written summaries of two of them, write summaries of the other two. Essays can be a part of courses and should be candidates for inclusion in portfolios.

6. *Summary courses.* Such courses are designed to cover and connect ideas from across the mathematics major. These may be specifically designed as summary courses and as such are usually called capstone courses, or they may be less specific, such as senior seminars or research seminars. Assessment of students performances in these courses provides good summary information about learning in the major.

7. *Oral presentations.* Oral presentations demonstrate speaking ability, confidence, and knowledge of subject matter. Students might be asked to prepare an oral presentation on a mathematics article. If these presentations are made in a summary course setting, then the discussion by the other students can serve both learning and assessment.

8. *Dialogue with students.* Student attitudes, expectations, and opinions can be sampled in a variety of ways and can be valuable in assessing learning. Some of the ways are student evaluations of courses, interviews by faculty members or administrators, advising interactions, seminars, student journals, and informal interactions. Also, in-depth interviews of individual students who have participated in academic projects as part of a group can provide insights into learning from the activities.

Student cooperation and involvement are essential to most assessment methods. When selecting methods appropriate to measuring student learning, faculty should exercise care so that all students are provided varied opportunities to show what they know and are able to do. The methods used should allow for alternative ways of presentation and response so that the diverse needs of all students are taken into account, while ensuring that uniform standards are supported. Students need to be aware of the goals and methods of the departmental assessment plan, the goals and objectives of the mathematics major and of each course in which they enroll, and the reason for each assessment measurement. In particular, if a portfolio of student work is collected, students should know what is going to go into those portfolios and why. Ideally, students should be able to articulate their progress toward meeting goals — in each course and in an exit essay at the end of the major.

Since some assessment measures may not affect the progress of individual students, motivation may be a problem. Some non-evaluative rewards may be necessary.

Step 4. Gather Assessment Data

After the assessment areas and methods are determined, the assessment is carried out and data documenting student learning are gathered. These data should provide answers to the second question in the introduction — how well are the students learning?

Careful record keeping is absolutely essential and should be well-planned, attempting to anticipate the future needs of assessment. Additional record storage space may be needed as well as use of a dedicated computer database. The data need to be evaluated relative to the learning goals and objectives. Evaluation of diverse data such as that in a student portfolio may not be easy and will require some inventiveness. Standards and criteria for evaluating data should be set and modified as better information becomes available, including longitudinal data gathered through tracking of majors through the degree program and after graduation. Furthermore, tracking records can provide a base for longitudinal comparison of information gathered in each pass through the assessment cycle.

Consistency in interpreting data, especially over periods of time, may be facilitated by assigning responsibility to a core group of departmental faculty members.

Ways to evaluate data include comparisons with goals and objectives and with preset benchmarks; comparisons over time; comparisons to national or regional norms; comparisons to faculty, student, and employer expectations; comparisons to data at similar institutions; and comparisons to data from other majors within the same institution.

If possible, students should be tracked from the time they apply for admission to long after graduation. Their interests at the time of application, their high school records, their personal expectations of the college years, their curricular and extracurricular records while in college, their advanced degrees, their employment, and their attitudes toward the institution and major should all be recorded. Only with such tracking can the long-term effectiveness of degree programs be documented. Comparisons with national data can be made with information from such sources as Cooperative Institutional Research Program's freshman survey data [7] and American College Testing's College Outcomes Measures project [8].

Step 5. Use the Assessment Results to Improve the Mathematics Major

The payoff of the assessment cycle comes when documentation of student learning and how it was achieved point the way for improvements for future students. Assessment should help guide education, so this final step in the cycle is to use the results of assessment to improve the next cycle. This is answering the third assessment question — what should be changed to improve learning? However, this important step should not be viewed solely as a periodic event. Ways to improve learning may become apparent at any point in the assessment cycle, and improvements should be implemented whenever the need is identified.

The central issue at this point is to determine valid inferences about student performances based on evidence gathered by the assessment. The evidence should show not only what the students have learned but what processes contributed to the learning. The faculty should become better informed because the data should reveal student learning in a multidimensional fashion.

When determining how to use the results of the assessment, faculty should consider a series of questions about the first four steps—setting goals and objectives, identifying learning and instructional strategies, selecting assessment methods, and documenting the results. The most critical questions are those about the learning strategies:

- Are the current strategies effective?
- What should be added to or subtracted from the strategies?
- What changes in curriculum and instruction are needed?

Secondly, questions should be raised about the assessment methods:

- Are the assessment methods effectively measuring the important learning of all students?
- Are more or different methods needed?

Finally, before beginning the assessment cycle again, the assessment process itself should be reviewed:

- Are the goals and objectives realistic, focused, and well-formulated?
- Are the results documented so that the valid inferences are clear?
- What changes in record-keeping will enhance the longitudinal aspects of the data?

IV. Resources for creating an effective assessment program

There are several resources available to help faculty and departments create an appropriate assessment program. *Assessment Practices in Undergraduate Mathematics* [9] contains over seventy case studies of assessment in mathematical sciences departments. A second volume of case studies, *Supporting Assessment in Undergraduate Mathematics* [10], emerged from an NSF-funded MAA project that included faculty development workshops in assessment. Results of this project can be found on line at www.maa.org/saum/. Departmental and institutional contexts for assessment are discussed in *CUPM Curriculum Guide 2004* [19], while several other resources for assessment are contained in this Guide's on-line *Illustrative Resources* found at www.maa.org/cupm/illres_refs.html.

One of the most important recent resources for assessment is the expanded edition of *How People Learn* [11], a summary of two major studies by National Research Council committees of research on what it means to know, from neural processes to cultural influences, and how to bring these results to bear on classroom practices. Pellagrino, *et al.* [12] builds on these insights of *How People Learn* to bring together advances in assessment and the understanding of human learning. Banta [13] provides an historical perspective, arguing that assessment is best seen as a reflective scholarly activity. Challis *et al.* [14] offer a practical resource from a British perspective, as does Houston [15]. In the U.S., Travers [16] contains the results of a major project to develop "indicators of quality" in undergraduate mathematics; it is available on CD as well as online. A subsequent synthesis report from this project [17] identifies questions and related statistics that form a "web of definitions" useful for describing the status and direction of a mathematics department's program. Connections between assessment, accreditation, articulation, and accountability are outlined in Ewell and Steen [18]. Finally, a report of the Association for Institutional Research (AIR) offers further in-depth case studies [20].

V. Conclusion

During an effective assessment cycle, students become more actively engaged in learning, faculty engage in serious dialogue about student learning, interaction between students and faculty increases and becomes more open, and faculty build a stronger sense of responsibility for student learning. All members of the academic community become more conscious of and involved in the way the institution works and meets its mission.

References

[1] *Guidelines for Programs and Departments in Undergraduate Mathematical Sciences.* Mathematical Association of America, Washington, DC, 1993.

[2] *Measuring What Counts.* Mathematical Sciences Education Board, National Research Council, National Academy Press, Washington, DC, 1993.

[3] *Assessment Standards for School Mathematics.* National Council of Teachers of Mathematics, Reston, VA, 1995.

[4]   *Principles of Good Practice for Assessing Student Learning*. American Association for Higher Education, Washington, DC, 1992.

[5]   "Mandated Assessment of Educational Outcomes." *Academe*, November-December, 1990, pp. 34–40.

[6]   Steen, Lynn Arthur, editor. *Heeding the Call for Change*. The Mathematical Association of America, Washington, DC, 1992.

[7]   Astin, A.W., Green, K.C., and Korn, W.S. *The American Freshman: Twenty Year Trend*. Cooperative Institutional Research Program, American Council on Education, University of California, Los Angeles, 1987. (Also annual reports on the national norms of the college freshman class.)

[8]   *College Level Assessment and Survey Services*, The American College Testing Program, Iowa City, 1990.

[9]   Gold, Bonnie, Sandra Z. Keith, and William A. Marion, editors. *Assessment Practices in Undergraduate Mathematics*. Mathematical Association of America, Washington, DC, 1999.

[10]   Steen, Lynn Arthur, et al., editors. *Supporting Assessment in Undergraduate Mathematics*. Mathematical Association of America, Washington, DC, 2005.

[11]   Bransford, J. D., Brown, A. L., & Cocking, R. R., editors. *How People Learn: Brain, Mind, Experience, and School: Expanded Edition*. National Academy Press, Washington, DC, 2000.

[12]   Pellegrino, J. W., Chudowsky, N. & Glaser, R. (Eds.). *Knowing What Students Know: The Science and Design of Educational Assessment*. Washington, DC: National Academy Press, Washington, DC, 2001.

[13]   Banta, Trudy W. and Associates. *Building a Scholarship of Assessment*. Jossey-Bass, San Francisco, 2002.

[14]   Challis, Neil, Ken Houston, and David Stirling. *Supporting Good Practice in Assessment in Mathematics, Statistics and Operational Research*. University of Birmingham, 2003.
        URL: mathstore.ac.uk/workshops/assess2003/challis.pdf.

[15]   Houston, Ken. "Assessing Undergraduate Mathematics Students." In *The Teaching and Learning of Mathematics at the University Level*, Derek Holton, Editor. Kluwer Academic Publishers, Dordrecht, 2001, pp. 407-422.

[16]   Travers, Kenneth J. *et al. Indicators of Quality in Undergraduate Mathematics*. University of Illinois Office for Mathematics, Science, and Technology Education, Urbana-Champaign, Illinois, 2002.
        URL: www.mste.uiuc.edu/indicators/.

[17]   Travers, Kenneth, *et al. Charting the Course: Developing Statistical Indicators of the Quality of Undergraduate Mathematics Education*. American Educational Research Association and the Office of Mathematics, Science and Technology www.mste.uiuc.edu/indicators/. Education (MSTE), University of Illinois, 2003.

[18]   Ewell, Peter T. and Lynn Arthur Steen. "The Four A's: Accountability, Accreditation, Assessment, and Articulation." *Focus: The Newsletter of the Mathematical Association of America*, 23:5 (May/June, 2003), p. 6–8.
        URL: www.maa.org/features/fouras.html.

[19]   *Undergraduate Programs and Courses in the Mathematical Sciences: CUPM Curriculum Guide 2004*. Mathematical Association of America, Washington, DC, 2004.

[20]   Madison, Bernard L., editor. *Assessment of Learning in Collegiate Mathematics: Toward Improved Courses and Programs*. Association for Institutional Research, Tallahassee, FL, forthcoming.